Microsoft Office PowerPoint 2007 A Beginners Guide

By WR Mills

AuthorHouse™
1663 Liberty Drive
Bloomington, IN 47403
www.authorhouse.com
Phone: 1-800-839-8640

©2009 W. R. Mills. All rights reserved.

No part of this book may be reproduced, stored in a retrieval system, or transmitted by any means without the written permission of the author.

First published by AuthorHouse 1/13/2010

IBSN: 978-1-4490-3236-4 (e)
ISBN: 978-1-4490-3235-7 (sc)

Library of Congress Control Number: 2010900401

Printed in the United States of America
Bloomington, Indiana

This book is printed on acid-free paper.

Microsoft Office PowerPoint 2007 A Beginners Guide

A training book for Microsoft PowerPoint 2007

By WR Mills

About the Author

Bill has a background in electronics and technology. He started writing software in 1982 and has expanded his programming skills to include C, C++, and Visual Basic. Bill also designs web sites. He designed a computer based telephone system for the hotel/motel market.

In 2007 he started teaching computer training classes and seems to have a knack for explaining things in a simple way that the average user can understand.

Bill is self-employed and lives in Branson Missouri with his wife Rose. They have three children, two sons and a daughter.

Preface

In 2007 I started teaching computer training classes. I was shocked at how much trouble the students had trying to understand the textbooks. I spent all of my time explaining what the textbook was trying to get across to the reader. It wasn't until I started getting ready for teaching the Microsoft Office 2007 series of classes that I finally gave up and started writing the textbooks myself.

These books are easy to understand and have step by step, easy to follow, directions. These books are not designed for the computer geek; they are designed for the normal everyday user.

It seems I have a knack for explaining things in a simple way that the average user can understand. I hope this book will be of help to you.

William R. Mills

Foreword

Dear Bill, I wanted to write a note of appreciation to you for your books: Microsoft Office Word 2007, Microsoft Office Excel 2007, and Microsoft Office PowerPoint 2007. I've used them all and found each one to be easy to read and very user friendly. If anyone needs to learn one of the 3 programs, but is even a little intimated, I strongly suggest they try one of your books. It's almost as good as taking a class with you as the instructor. If I didn't understand a step, I just went back to the pervious step and tried it again --- and it always worked!! There's just enough humor in the text to keep the reading interesting; never dull, but fun and light. Just what a beginner needs. Again, Bill, I thank you for creating these books that make learning something I needed to learn fun and easy. Sincerely, Cyndy O

Important Notice:

There will be times during this book that you be asked to open a specific file for the lesson. These files can be downloaded from the EZ 2 Understand Computer Books web site.

Open your internet browser (probably Internet Explorer) and go to www.ez2understandcomputerbooks.com. Click your mouse on the <u>Lesson Files</u> link toward the top. This will take you to the page where you can download the files needed. There are directions on the page to help you with the download. They are repeated below for your convenience.

To download the files follow the following steps:

1) Right-click your mouse on the file(s) you want. These are zipped files and contain the lessons that you will need for each book.

2) Select the "Save Target As" choice. Make sure the download is pointed to a place on your hard drive where you can find it, such as My Documents.

3) Click the Save button

4) The files are zipped files and will need to be extracted to access the contents. To extract the files right-click on the file and choose "Extract All".

Table of Contents

Chapter One	The Basics	1
Lesson 1 - 1	Starting PowerPoint	2
Lesson 1 – 2	Understanding the PowerPoint Screen	3
Lesson 1 – 3	The Office Button	5
Lesson 1 – 4	The Ribbon – An Overview	12
Lesson 1 – 5	The Quick Access Toolbar	15
Lesson 1 – 6	Using the Keyboard	18
Lesson 1 – 7	Creating a Presentation	25
Lesson 1 – 8	Saving the Presentation	32
Lesson 1 – 9	Closing the Presentation	37
Lesson 1 – 10	Opening a Presentation	38
Lesson 1 – 11	Printing a Presentation	41
Chapter One	Review	48
Chapter One	Quiz	49
Chapter Two	The Ribbon – A Closer Look	50
Lesson 2 – 1	The Home Tab	51
Lesson 2 – 2	The Insert Tab	55
Lesson 2 – 3	The Design Tab	57
Lesson 2 – 4	The Animations Tab	58
Lesson 2 – 5	The Slide Show Tab	59
Lesson 2 – 6	The Review Tab	60
Lesson 2 – 7	The View Tab	61
Lesson 2 – 8	The Developer Tab	64
Chapter Two	Review	65
Chapter Two	Quiz	66
Chapter Three	Editing a Presentation	67
Lesson 3 – 1	Inserting Slides	68
Lesson 3 – 2	Moving Slides	70
Lesson 3 – 3	Editing Slides	71
Lesson 3 – 4	Adding Text	74
Lesson 3 – 5	Selecting, Replacing, and Deleting Text	76

Lesson 3 – 6	Using Cut, Copy, and Paste	78
Lesson 3 – 7	Using Undo and Redo	80
Lesson 3 – 8	Using Spell Check	82
Lesson 3 – 9	Finding Information	87
Lesson 3 – 10	Adding Comments	90
Chapter Three	Review	93
Chapter Three	Quiz	94
Chapter Four	Managing Files	95
Lesson 4 – 1	Making New Folders	96
Lesson 4 – 2	Moving Files	99
Lesson 4 – 3	Copying Files	101
Chapter Four	Review	103
Chapter Four	Quiz	104
Chapter Five	Formatting	105
Lesson 5 – 1	Formatting Text using the Ribbon	106
Lesson 5 – 2	Using the Dialog Box	111
Lesson 5 – 3	Number and Bullet Lists	113
Lesson 5 – 4	Paragraph Alignment	117
Lesson 5 – 5	Using the Format Painter	120
Chapter Five	Review	122
Chapter Five	Quiz	123
Chapter Six	Using Themes and Styles	124
Lesson 6 – 1	Applying a Theme	125
Lesson 6 – 2	Modifying a Theme	127
Lesson 6 – 3	Adding Background Styles	130
Chapter Six	Review	133
Chapter Six	Quiz	134
Chapter Seven	Working with Tables	135
Lesson 7 – 1	Creating a Table	136
Lesson 7 – 2	The Table Tools Design Tab	141
Lesson 7 – 3	The Table Tools Layout Tab	145
Chapter Seven	Review	155

Chapter Seven	Quiz	156
Chapter Eight	Working with Graphics	157
Lesson 8 – 1	Inserting Pictures	158
Lesson 8 – 2	Working with Text Boxes	162
Lesson 8 – 3	Working with ClipArt	166
Lesson 8 – 4	Positioning Pictures	170
Lesson 8 – 5	Using the Picture Tools	172
Lesson 8 – 6	Working with Charts	179
Chapter Eight	Review	190
Chapter Eight	Quiz	191
Chapter Nine	Transitions & Animation	192
Lesson 9 – 1	Slide Transitions	193
Lesson 9 – 2	Transitions Sounds	195
Lesson 9 – 3	Animations	198
Chapter Nine	Review	206
Chapter Nine	Quiz	207
Chapter Ten	Multimedia	208
Lesson 10 – 1	Adding Audio Clips	209
Lesson 10 – 2	Adding Video Clips	211
Chapter Ten	Review	213
Chapter Ten	Quiz	214
Chapter Eleven	Presenting the Presentation	215
Lesson 11 – 1	Delivering on a Computer	216
Lesson 11 – 2	Rehearsing Timings	219
Lesson 11 – 3	Self Playing Presentation	223
Chapter Eleven	Review	227
Chapter Eleven	Quiz	228
Chapter Twelve	Protecting your Presentation	229
Lesson 12 – 1	Creating a Password to open the Presentation	230
Lesson 12 – 2	Creating a Password to Edit a Presentation	233
Chapter Twelve	Review	237
Chapter Twelve	Quiz	238

Chapter One The Basics

Microsoft PowerPoint 2007 is a powerful desktop presentation program. PowerPoint will let you turn your ideas into slide shows and presentations. PowerPoint has a variety of tools for the user. These tools will allow the user to produce professional presentations. With all of these tools available to you, PowerPoint 2007 is extremely easy to use.

The first thing that you are going to notice is that PowerPoint 2007 looks different than any other version of PowerPoint. This is because of the new user interface. You might ask why this version of PowerPoint is better than the version you are use to. Let me answer that in this way; do you remember searching through a series of menus and submenus to find a command? That is all a thing of the past. PowerPoint 2007 has the Ribbon. Wow! Are you excited yet?

Is the Ribbon scary? Probably. Is it intimidating? More than likely. Is it better and easier to use? Yes definitely. The Ribbon is based more on how people actually use their computer.

The Ribbon is divided into Task Orientated Tabs. Each tab has groups of related commands. Everything you need is right at your fingertips. You will not have to search through menus and submenus until you want to pull your hair out, trying to find a command.

PowerPoint can automatically check your spelling and grammar, and it can correct common mistakes. For example if you type hte PowerPoint will automatically change it to the. With PowerPoint you can insert charts, tables, and pictures into your presentations.

There's also the new Quick Styles. These are ready-made styles that give your presentation a professional makeover fast.

This chapter is an introduction to the basics of Microsoft PowerPoint. We will cover what you need to create, print, and save a presentation. We will also start covering the new Ribbon.

Lesson 1 - 1   Starting PowerPoint

Starting PowerPoint is a simple process. The first thing you have to do is have your computer on and the desktop showing. I know, I didn't have to say that, but I did, too late to take it back now. The PowerPoint program is located in the Microsoft Office folder.

Click on the Start Button

Select All Programs

Move the mouse to Microsoft Office and click on Microsoft Office PowerPoint 2007 **from the menu that slides out to the right side**

In a few moments, PowerPoint will appear on your screen. It should look similar to Figure 1-1. You may not have all of the options at the top; it depends on the available screen width of your monitor.

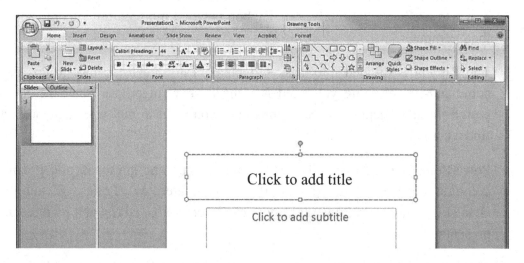

Figure 1-1

From here we can start creating our presentation. Before we jump headlong into all of the different things that you can do, let's get use to the screen.

Lesson 1 – 2   Understanding the PowerPoint Screen

The moment you start PowerPoint 2007 you will notice some major changes. Microsoft completely redesigned the user interface. Microsoft pretty much went back to the drawing board to design the way you use PowerPoint. Now it works now is based on how most people actually use the program. Figure 1-2 shows the PowerPoint screen.

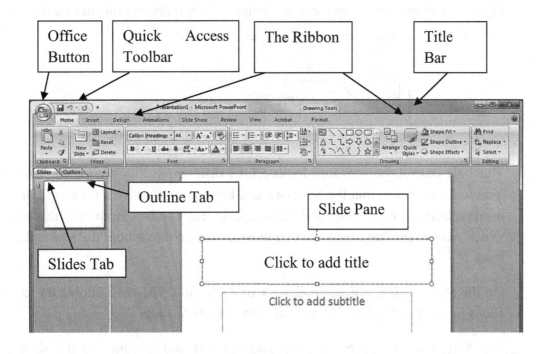

Figure 1-2

Figure 1-3 shows a close-up view of the Office Button. The Office button is actually located on the end of the Quick Access Toolbar. We will discuss the Office button more in the next lesson.

Figure 1-3

Figure 1-4 shows the Quick Access Toolbar. We will discuss this in Lesson 1-5.

Figure 1-4

The Title Bar (at the very top) looks a little different in this version of PowerPoint. You will also notice that the Standard and Formatting Toolbars are gone. They have been replaced by the Ribbon and the Quick Access Toolbar. Both the Ribbon and the Quick Access Toolbar are new and will be discussed throughout the remainder of the book.

On the Slides Tab is a thumbnail view of each slide. This will allow you to see the slides in order without having to switch from slide to slide.

The Slide Pane is where you will add the text and graphics to the slide in the presentation. This is also where you will add the formatting.

Lesson 1 – 3   The Office Button

The Office Button, for the most part, has taken the place of the old File section of the menu bar. As you can see the menu bar does not exist in this version of PowerPoint. In this lesson we will examine the Office Button and see just how it works.

Using your mouse, click on the Office Button

When you click on the Office Button, a menu will drop down giving you several choices of things you are able to do. This is shown in Figure 1-5.

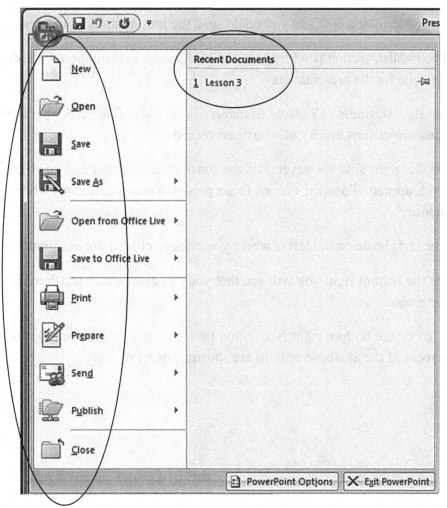

Figure 1-5

On the left side you will notice that many of the choices were the same as when you clicked on the file button of the older style menu bar. You can start a new presentation or open an existing presentation. You also have the Save and Save As choices. You may notice that the drop-down choices are divided into two sections. The most popular choices are put at the top. The lesser used options are placed toward the bottom.

If you needed to print the presentation, you would find the printer options under the print choice.

The Prepare section is where you would look at the properties of the presentation as well as encrypt it so no one could open it or edit it without a password.

The Send option is where you could send the presentation as an e-mail or a fax.

The Publish section is where you would share the presentation with others or create a new site for the presentation.

In the Business Contact Manager, you can link this presentation to the communications history of a business record.

On the right side are several of the most recent PowerPoint presentations that you have opened. To open one of these presentations you simply click on it with the mouse.

The last choice on the left is where you click to close a presentation.

On the bottom right you will see that you can also exit PowerPoint from this part of the menu.

Also on the bottom right is a button to access the PowerPoint Options. The Option screens of the available options are shown starting in Figure 1-6.

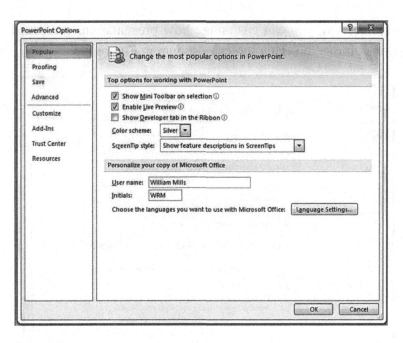

Figure 1-6

From this screen you can change the most popular options, such as should the Mini Toolbar be shown when you select text, and what user name to attach to a document.

In the Proofing section, shown in Figure 1-7, you can make changes as to how PowerPoint corrects and formats words as you type. You can also choose if PowerPoint will ignore certain words.

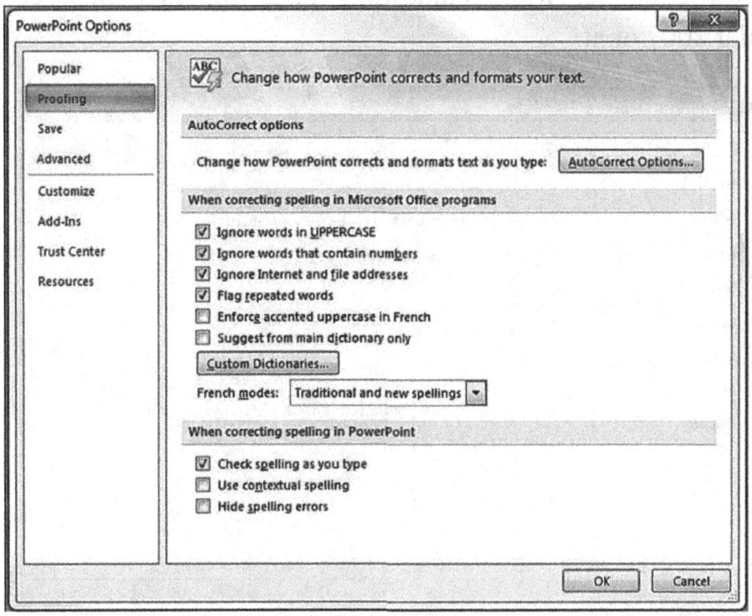

Figure 1-7

In the Save section, you can customize how presentations are saved and if the auto save option is turned on. This screen is shown in Figure 1-8.

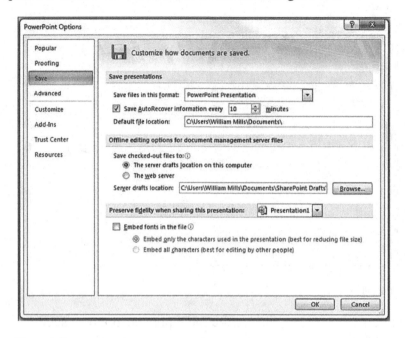

Figure 1-8

In the Typography section you can set which characters cannot start or end a line. Figure 1-9 shows the Typography section. Not everyone will have this section. I am running PowerPoint 2007 on a machine with Microsoft XP on it and it has this section. My computer with Windows 7 does not have this section when I run PowerPoint 2007 on it.

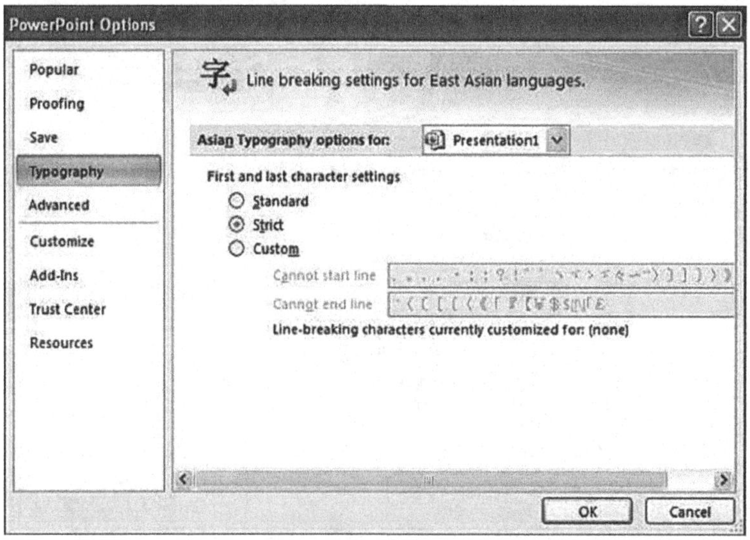

Figure 1-9

In the Advanced section there are some options for working with PowerPoint (shown in Figure 1-10). The options include such things as: Editing options like the maximum number of times you can use the undo feature and is the drag and drop option available. There are also options for the slide show and various print options.

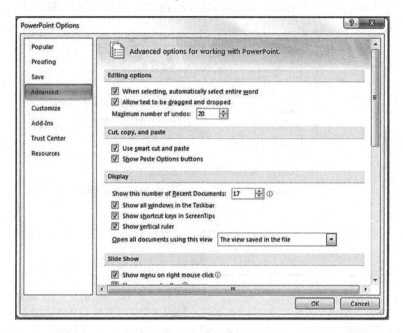

Figure 1-10

The Customize section allows you to add and remove icons from the Quick Access Toolbar. You can also move the toolbar to below the Ribbon instead of having it above the Ribbon. See Figure 1-11 for the Customize screen.

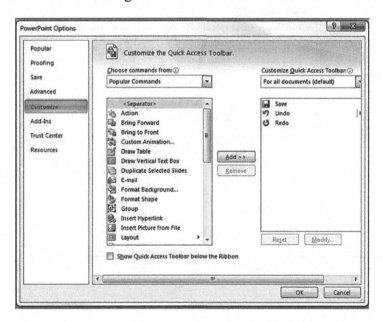

Figure 1-11

The add-ins section shows the "extra" things that have been added to help PowerPoint work better. This is shown in Figure 1-12.

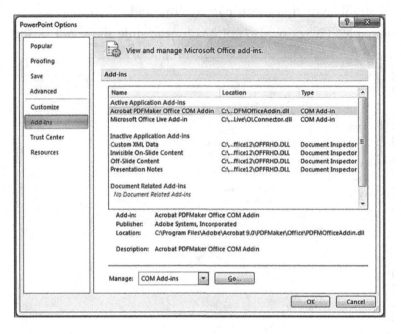

Figure 1-12

The Trust Center contains security and privacy settings. Microsoft recommends that you do not change these settings. See Figure 1-13 for the Trust Center.

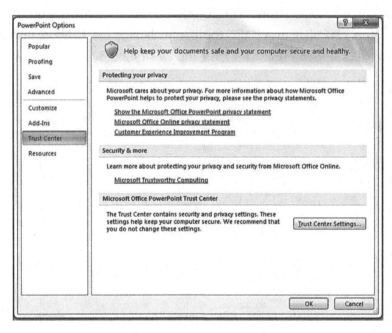

Figure 1-13

The last section of the PowerPoint Options is the Resource Center. From here you can get updates, contact Microsoft, etc. This screen is shown in Figure 1-14.

Figure 1-14

Lesson 1 – 4   The Ribbon – An Overview

The Ribbon has been designed to offer an easy access to the commands that you (the user) use most often. You no longer have to search for a command embedded in a series of menus and submenus. The Ribbon has a series of Tabs and each tab is divided into several groups of related commands. Figure 1-15 shows the Ribbon across the top of the PowerPoint program.

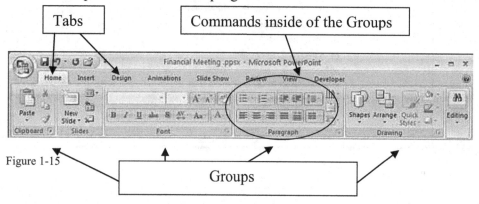

Figure 1-15

There are three major components to the Ribbon.

Tabs:

There are eight basic tabs across the top.

> The Home Tab contains the commands that you use most often.
>
> The Insert Tab contains all of the objects that can be inserted into a Presentation.
>
> The Design Tab contains the choices for how each page will look.
>
> The Animations Tab contains things such as how you transition from one slide to the next slide.
>
> The Slide Show Tab contains such things as setting up the slide show and recording narrations.
>
> The Review Tab has things that are related to proofing and adding comments.
>
> The View Tab allows you to change to the different views that are available.
>
> The Developer Tab contains programming and controls that can be added to the slides.

Groups:

Each Tab has several Groups that show related item together.

Look at the Home Tab to see an example of the related Groups.

The Home Tab has the following Groups: Clipboard, Slides, Font, Paragraph, Drawing, and Editing.

Commands:

A Command is a button, a box to enter information, or a menu.

The Clipboard Group, for example, has the following commands in it: Cut, Copy, Paste, and Format Painter.

When you first glance at a group, you may not see a command that was available from the menus of the previous versions of PowerPoint. If this is the case you need not worry. Some Groups have a small box with an arrow in the lower right side of the Group. See figure 1-16 for a view of a group with this arrow.

Figure 1-16

This small arrow is called the Dialog Box Launcher. If you click on it, you will see more options related to that Group. These options will usually appear in the form of a Dialog Box. You will probably recognize the dialog box from previous versions of PowerPoint. These options may also appear in the form of a task pane. Figure 1-17 shows the Font Dialog Box.

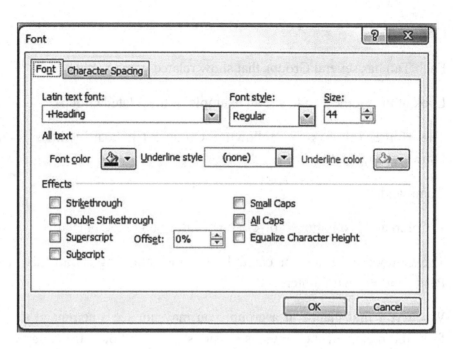

Figure 1-17

Speaking of previous versions, if you are wondering whether you can get the look and feel of the older versions of PowerPoint back, the answer is simple, **no you can't**.

The good news is that after playing with and using the Ribbon, you will probably like it even better. It really does make working with PowerPoint easier. The Ribbon will be used extensively and the tabs will be covered in more detail later as we go through this book.

Lesson 1 – 5   The Quick Access Toolbar

The Ribbon, as you will find out, is wonderful, but what if you want some commands to always be right at your fingertips without having to go from one tab to another? Microsoft gave us a toolbar for just that purpose. This toolbar is called the Quick Access Toolbar and is located just above, or below, and to the left end of the Ribbon. The Quick Access Toolbar is shown in Figure 1-18.

Figure 1-18

The Quick Access Toolbar contains the Office button, the Save button, the Undo and Redo buttons. These are things that you normally use over and over and you will want them available all of the time.

There is even more good news, if you want to add an item to the toolbar, the process is very simple. At the right end of the toolbar is an arrow pointing downwards. If you click on this arrow, a new drop down menu will come onto the screen, as shown in Figure 1-19.

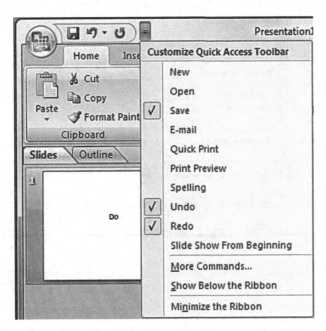

Figure 1-19

15

From this menu you can choose from the standard choices or you can customize the toolbar to suit your needs by clicking on the More Commands choice.

You can also choose to show the tool bar below the Ribbon instead of above it. I normally have my computer set to show the Quick Access Toolbar below the Ribbon, probably because the toolbars were always below the menu bar in the older versions.

If you want to add an item from the standard choices all you have to do is click on the item you want to add. The drop down menu will disappear and the new item will be added to the toolbar.

Add the following choices to the Quick Access Toolbar: Open, New, Spelling, Quick-Print, and Print Preview

If the option you want to add is not listed in the standard choices, all of the available options are listed under the More Commands.

Figure 1-20 shows the PowerPoint Options Dialog box that will come to the screen if you choose the More Commands option.

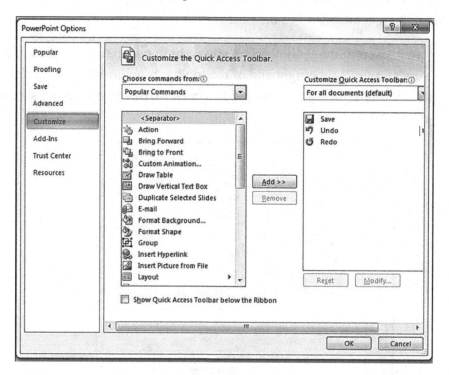

Figure 1-20

If you wish to add an item to the Quick Access Toolbar, all you need to do is click on the option on the left and then click the Add button in the center. When you are finished adding items, click the OK button to place them in the toolbar. You will probably find that there are several things that you will use over and over with every presentation and you will want to place them in the Quick Access Toolbar just because this will save you so much time.

Lesson 1 – 6   Using the Keyboard

What about all of you people who prefer to use the Keyboard over the mouse? I have not forgotten about you, and this lesson is just for you.

As you have more than likely noticed the old menu bars are not there anymore. Before you break down and the tears start to roll, let's see what we can do.

Microsoft gave us some options for the keyboard user. Although the menus are not there, you can use the keyboard to access the different parts of the Ribbon. Not only can you access the Ribbon, but the old shortcuts (using the CTRL key) that you are use to using are still there and still working.

Below are tables showing some of the available keyboard shortcuts. A complete list of all of the shortcuts is provided for you on your computer in the Help section. To find the list click on the Help button (the small question mark in the upper right corner of the screen) and type keyboard shortcuts in the search window. Some of that list has reproduced for you in the tables below.

To do this	Press
Open the Help window.	F1
Close the Help window.	ALT+F4
Switch between the Help window and the active program.	ALT+TAB
Go back to **PowerPoint** Home.	ALT+HOME
Select the next item in the Help window.	TAB
Select the previous item in the Help window.	SHIFT+TAB
Perform the action for the selected item.	ENTER

In the **Browse PowerPoint Help** section of the Help window, expand or collapse the selected item, respectively.	ENTER
Move back to the previous Help topic (**Back** button).	ALT+LEFT ARROW or BACKSPACE
Move forward to the next Help topic (**Forward** button).	ALT+RIGHT ARROW
Scroll small amounts up or down, respectively, within the currently displayed Help topic.	UP ARROW, DOWN ARROW
Scroll larger amounts up or down, respectively, within the currently displayed Help topic.	PAGE UP, PAGE DOWN
Display a menu of commands for the Help window. This requires that the Help window have the active focus (click in the Help window).	SHIFT+F10
Stop the last action (**Stop** button).	ESC
Refresh the window (**Refresh** button).	F5
Print the current Help topic. NOTE If the cursor is not in the current Help topic, press F6 and then press CTRL+P.	CTRL+P

To do this	Press
Change the font.	CTRL+SHIFT+F
Change the font size.	CTRL+SHIFT+P
Increase the font size of the selected text.	CTRL+SHIFT+>
Decrease the font size of the selected text.	CTRL+SHIFT+<
Move clockwise among panes in Normal view.	F6
Move counterclockwise among panes in Normal view.	SHIFT+F6
Switch between **Slides** and **Outline** tabs in the Outline and Slides pane in Normal view.	CTRL+SHIFT+TAB
Open the **Font** dialog box to change the font.	CTRL+SHIFT+F
Open the **Font** dialog box to change the font size.	CTRL+SHIFT+P
Increase the font size.	CTRL+SHIFT+>
Decrease the font size.	CTRL+SHIFT+<
Start the presentation from the beginning.	F5
Perform the next animation or advance to the next slide.	N, ENTER, PAGE DOWN, RIGHT ARROW, DOWN ARROW, or SPACEBAR
Perform the previous animation or return to the previous slide.	P, PAGE UP, LEFT ARROW, UP ARROW, or BACKSPACE

Go to slide *number*.	*number*+ENTER
Display a blank black slide, or return to the presentation from a blank black slide.	B or PERIOD
Display a blank white slide, or return to the presentation from a blank white slide.	W or COMMA
Stop or restart an automatic presentation.	S
End a presentation.	ESC or HYPHEN
Erase on-screen annotations.	E
Go to the next slide, if the next slide is hidden.	H
Set new timings while rehearsing.	T
Use original timings while rehearsing.	O
Use a mouse click to advance while rehearsing.	M
Return to the first slide.	1+ENTER
Redisplay hidden pointer or change the pointer to a pen.	CTRL+P
Redisplay hidden pointer or change the pointer to an arrow.	CTRL+A

Hide the pointer and navigation button immediately.	CTRL+H
Hide the pointer and navigation button in 15 seconds.	CTRL+U
Display the shortcut menu.	SHIFT+F10
Go to the first or next hyperlink on a slide.	TAB
Go to the last or previous hyperlink on a slide.	SHIFT+TAB
Perform the "mouse click" behavior of the selected hyperlink.	ENTER while a hyperlink is selected

As you can see there are several pages of shortcut keys available for you to use. This is not even a complete list of all of the shortcut keys that are available. If you still need more commands, click on the help button and search for Keyboard Shortcuts. The Help button is the small question box at the top right side of the screen.

Are you ready for even more good news? Microsoft has included new shortcuts with the Ribbon. Why you might ask. It is because this change brings two major advantages. First there are shortcuts for every single tab on the Ribbon and second because the many of the shortcuts require fewer keys.

The new shortcuts also have a new name: **Key Tips**

Using the keyboard press the Alt **key**

Pressing the Alt key will cause the **Key Tip Badges** to appear for all Ribbon tabs, the Quick Access Toolbar commands, and the Microsoft Office Button. After the Key Tip Badges appear, you can press the corresponding letter or number on the badge for the tab or the command you want to use. As an example, if you pressed Alt and then H you would bring the Home tab to the front. Figure 1-21 shows what the Ribbon looks like after pressing the Alt key.

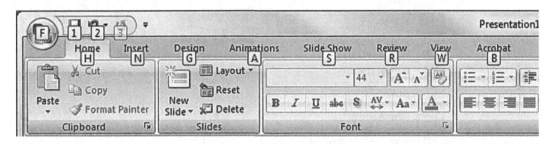

Figure 1-21

Note: You can still use the old Alt + shortcuts that accessed the menus and commands in the previous versions of PowerPoint, but because the old menus are not available, you will have no screen reminders of what letters to press, so you will need to know the full shortcut to be able to use them.

But that is not all of the shortcuts that are available. Microsoft has included shortcut menus. These are menus that you can access by right-clicking on an object or text. A shortcut menu is shown in Figure 1-22.

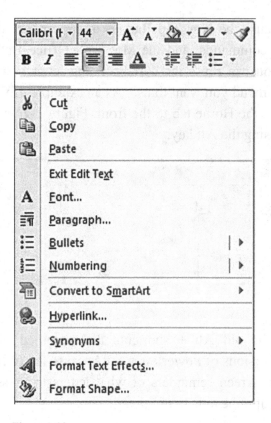

Figure 1-22

This shortcut menu shows all of the things that are available for you to do to the selected object. The bottom part looks a lot like the Edit from the previous version's menu bar. You will also notice the different formatting options that you can perform from the shortcut menu. See I told you that this version was cool!

Lesson 1 – 7 Creating a Presentation

When you first start PowerPoint 2007 you should have a blank screen like the one shown in Figure 1-23.

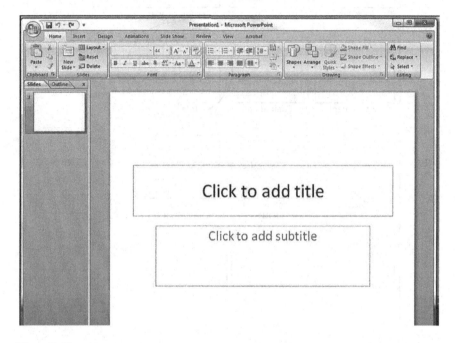

Figure 1-23

The first thing we will want to do is decide what type of slide we need on the screen. The default slide is shown on the screen, but this may not be what is required for your presentation. Normally this will work for the first slide because it has a place for a title and a sub-title. If, on the other hand this will not work, we can change this slide to a different type.

On the Home Tab there is a group called the Slides Group. In this group is a command called Layout. Clicking this command will bring a drop down menu to the screen which will allow us to change the layout of the current slide.

Click the Layout **command**

The result of clicking this command is shown in Figure 1-24.

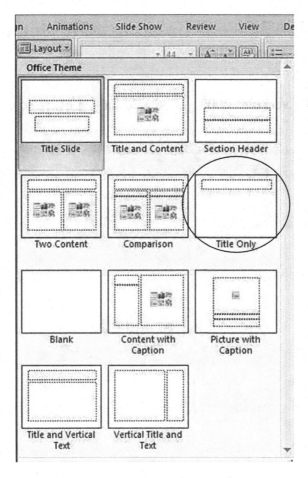

Figure 1-24

As you can see there are several different types of layouts. If we wanted to change the type all we would have to do is click on the layout we need and the slide would immediately change to that type of slide. If we wanted a title only we would click on the slide that has title only under it.

Click on slide that says Title Only

Note: To leave the default slide as the current type you could have just clicked outside of the drop down menu.

At this point we should add a title to our presentation.

Click the mouse where it says "Click to add title"

The text that is inside the box will disappear and the insertion point (the flashing vertical line) will take its place. All we now have to do is start typing the title.

Using the keyboard type the following

Financial Meeting of the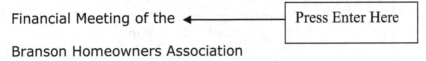

Branson Homeowners Association

The slide should look like Figure 1-25.

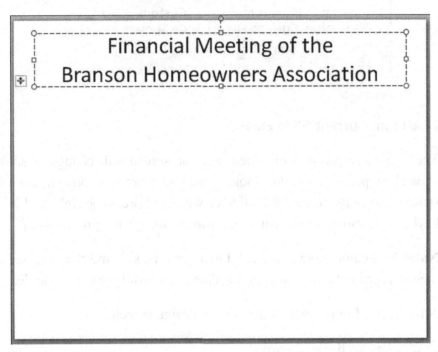

Figure 1-25

The first slide in our presentation is finished. That wasn't so bad, was it? Want to see what it will look like when we run the presentation?

On the Slide Show Tab and in the Start Slide Show Group is the command to start the slide show from the beginning or the current slide. It won't matter which one we click on at this time because we only have one slide. Since we want to see what this slide will look like, let's use the from current slide choice (see Figure 1-26).

Figure 1-26

Click on the From Current Slide **choice**

The normal screen will disappear and the screen will change to show you how the slide show presentation will look. If we had more than one slide we could press the spacebar to move to the next slide or we could press the right and left arrows on the keyboard to move forward and backwards through the presentation.

Note: You could have also held down the shift key and then pressed the F5 key and then released both keys to start the slide show from the current slide.

Now we need to go back to the normal design screen.

Press the ESC **key on the keyboard**

The ESC key is on the top left of the keyboard.

Well the first slide is finished but we need another slide. After all what good is a presentation with only one slide? There are a couple of ways to add another slide to our presentation. Let's do it the hard way first. There is no hard way, I was only kidding, but there is more than one way to add a slide. We can add the default type of slide and then change it if it does not meet or requirements or we can choose the type of slide right off the bat. Let's choose the type of slide we want and get it over with (if the default type was incorrect we would have to change the layout anyway).

On the Home tab and in the Slides Group is a button for adding a new slide. It has the name New Slide on it (good thinking Microsoft). If you click on the top part of the command, the default slide layout will be added to your presentation. If you click the down arrow, you get to choose the slide layout from the menu.

Click the down arrow on the New Slide command

A menu of the slide layouts (which should look familiar) will drop down and allow you to choose the layout that will fit your needs. Our second slide is going to have more text on it, so we will need to find a layout that has room for more than one line of text. Figure 1-27 shows the drop down menu.

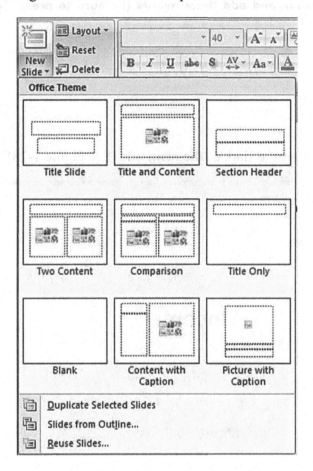

Figure 1-27

Choose the Title Slide **by clicking on it with the mouse**

A new slide will jump into your presentation and it will look similar to the one we had before. Using this slide we will add the text we want our viewers to see on this slide.

Click inside the textbox that has "Click to add title**" on it and then type**

> Income

Click outside the textbox to see what the slide now looks like

The outline around the box will disappear and the word Income will remain. Now we will add what is to appear under the word income. We will put this in the textbox that has "Click to add subtitle" in it.

Click inside the subtitle textbox and add these words (be sure to press Enter after the first two topics)

Dues

Fines

Monthly Fees

The subtitle does not appear as dark as the main title (See Figure 1-28).

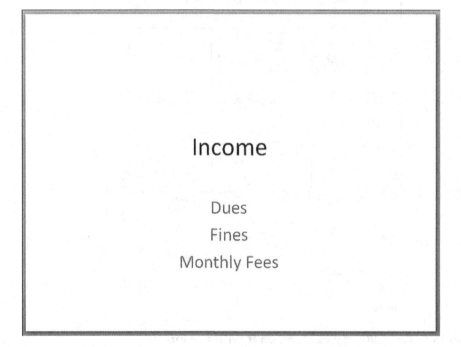

Figure 1-28

We are going to continue adding to this presentation a little later, but for now this will be just enough for the first lesson. Before we move to the next lesson let's see how the presentation will look when we run it.

On the Slide Show Tab and in the Start Slide Show Group click on the "From beginning" command

This will start the slide show from the beginning and we can see how we are doing. This is good for viewing the first slide, but what about the second slide?

Using the keyboard press the right arrow key one time

Pressing the right arrow key will allow us to move to the next slide, and you will be able to see slide two.

Using the keyboard press the right arrow key again

There is no third slide (not yet anyway) for us to view so the screen will go to a black screen with the words "End of slide show click to exit" at the top.

Click the mouse anywhere on the screen to go back to the PowerPoint screen

Do not close the PowerPoint program. Instead go directly to the next lesson

Lesson 1 – 8   Saving the Presentation

Now that you have gone through all of the trouble of making the slides, we need to save it so that nothing will happen to our Presentation. I don't know about you, but I have gone through situations like this only to have the power go off and everything is gone. Before that happens, lets learn how to save our presentation.

At this point I am going to make an assumption. The assumption is that the presentation from the previous lesson is still open. If it is not open, open it now.

Click the Office Button

When you click the Office button a drop down menu will appear as shown in Figure 1-29.

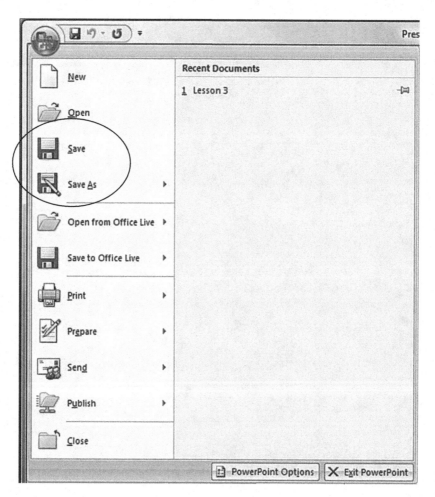

Figure 1-29

32

Note: The recent documents side of yours will not look like the one in figure 1-25; since these are my recent documents not yours.

You will notice that there are two save choices. Now would be a good time to explain the difference between the two of them. If this is the first time you have saved the current presentation, The Save choice will bring the Save As dialog box to the screen. If this is not the first time you have saved the presentation the choice will have different results.

Save: This will replace the existing presentation (file) with the newer version that is displayed on your screen. All changes made will be saved and the original presentation, before you made any changes, will be gone.

Save As: This will allow you to save the currently displayed presentation with a different name. This will allow you to keep the original presentation just as it was before any changes were made to it and the new version will also be saved only under a different name.

Move the mouse to the Save As choice

Another menu will slide out to the side.

The first thing you have to do is make an important decision: what format should I use to save this presentation? Figure 1-30 shows the choices you have to choose from.

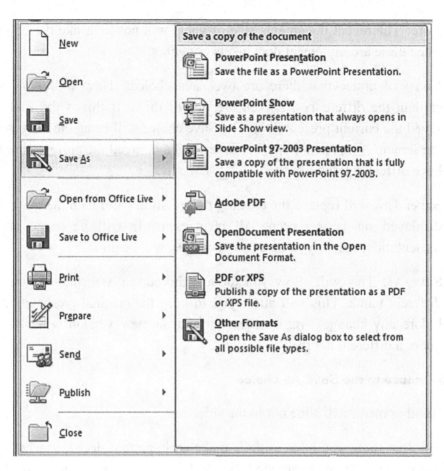

Figure 1-30

The two most obvious choices are: Save in the default format, which is PowerPoint 2007 and Save in the older format of PowerPoint 97 – 2003. If you were going to share this with someone who does not have PowerPoint 2007 on their computer but has an older version of PowerPoint, you would choose the PowerPoint 97-2003 workbook choice.

There is another choice you might want to consider. This is the PowerPoint Show choice. If you choose this choice anytime you open the presentation, it will open in the slide show view.

We are going to save our presentation in the default format, which is PowerPoint 2007.

Click on the top choice, PowerPoint presentation

The Save as dialog box will come on to the screen and is shown in Figure 1-31.

34

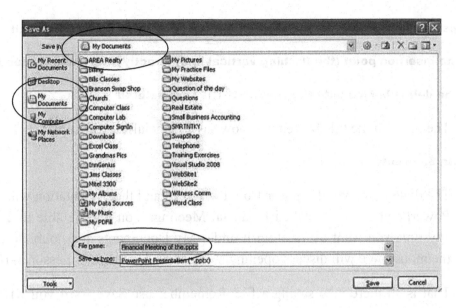

Figure 1-31

If this is the first time you have saved this presentation the "Save As" dialog box will also come to the screen. The reason for this is because the first time you save a presentation you have to give it a name. If it already has a name and you select the Save choice it will save the new version. I had you choose the Save As choice just to make sure you knew about the different versions available. If you had simply clicked the save choice, PowerPoint would have saved the presentation in the default format which is PowerPoint 2007.

The "Save As" dialog box appears asking where to save the Presentation and what name you want to give the presentation as shown in Figure 1-22. The default location is usually My Documents (unless you are already working in another folder), and the suggested name (usually the first line of the document) are already in place. If they are acceptable, simply click on "Save" to save the presentation. Your "Save As" screen may look different because these are my documents not yours.

Make sure My Documents **is selected on the right side (if it is not click on it with the mouse).**

Now all we have to do is give the presentation a unique name. It is essential that every file has a unique name. This will allow us to keep our computer organized and also allow us to find the file when we need it. The name should reflect something about the presentation, such as why it was created.

This presentation is being used for a slide show about the Financial Meeting of the Branson Homeowners Association, so a name like "Financial Meeting" is more practical than a name like Presentation 17.

Click inside the box that has the name "Financial Meeting of the.pptx" in it

Move the insertion point (the flashing vertical line) directly after the g in meeting

Press the delete button until the words "of the" are deleted

The name in the title box should now say Financial Meeting.pptx

Click the Save button

The dialog box will disappear from the screen and the presentation will be saved as a PowerPoint 2007 file called Financial Meeting. You will be able to access this file any time you want by locating it under My Documents and double clicking it with the mouse. We will discuss opening an existing document in Lesson 1-10.

That is all there is to saving a file. Remember use Save As if you wish to keep the original presentation as it was before any changes were made and use Save if you wish to replace the original file with the revised version.

Note: From now on, when you are asked to save a presentation, make sure you save it in the **MY Documents** folder until you are told differently.

Lesson 1 – 9   Closing the Presentation

Now that we have saved our work we can safely close the file. This will be a very short lesson, as there is not very much to closing a presentation (file).

Click on the Office Button

From here we will simply click on the Close choice at the bottom (see Figure 1-32).

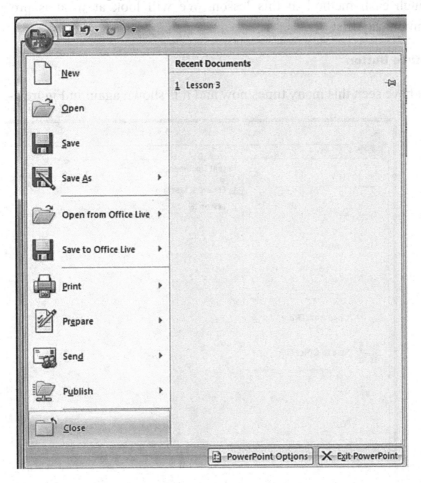

Figure 1-32

Click on the Close choice at the bottom

The presentation will close and you will end up with the main part of the screen being blank. In the next lesson we will see how to open the presentation.

Lesson 1 – 10   Opening a Presentation

If necessary open PowerPoint

Remember that PowerPoint is located in the Microsoft Office folder under all programs which is under the Start button.

There are three basic ways to open an existing PowerPoint presentation. We will go through each method in this lesson. We will look at what is probably the most common way first.

Click the Office Button

You have seen this many times now and it is shown again in Figure 1-33.

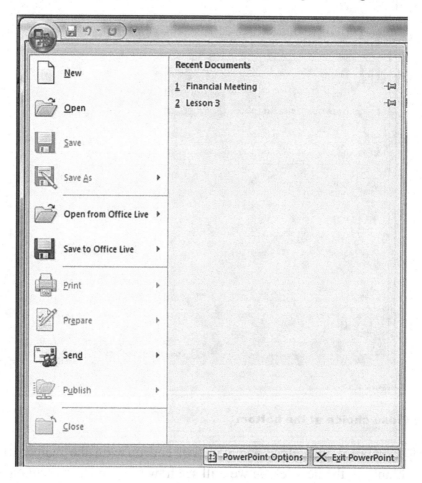

Figure 1-33

Click the Open Button

This will bring the Open Dialog Box to the screen as shown in Figure 1-34.

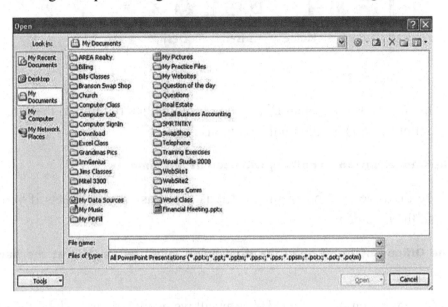

Figure 1-34

The normal starting place to look for a file is under My Documents, so that is the default place that should come up when you first see the Open dialog box. Since this is the place where we saved the file, you should see a file named Financial Meeting.pptx. You will notice that the Open button is faded out and you cannot click on it at this point. Right now the button is disabled. As soon as you click on a file name the Open button will change and be enabled and be available to be clicked.

Click on Financial Meeting.pptx **and then click the Open button**

As soon as you click the Open button, PowerPoint will start opening the Presentation. It may take a second or two depending on the speed of your computer for the screen to actually change. But in a few moments the presentation will whisk onto your computer screen and you will be able to work with it.

Close the presentation as we did in lesson 1-9

The second method for opening a document may be a little easier, but I am not sure that as many people use it as the first method we used.

Click on the Open button on the Quick Access Toolbar

The Quick Access Toolbar is show in Figure 1-35 and the open button is circled.

Figure 1-35

This will cause the Open Dialog box to immediately jump to the screen and you can continue just as we did on the previous page.

Click the Cancel button so nothing will open at this time

The third method for opening a file is about as easy as it gets if you have recently had the file open.

Click the Office Button to open the drop down menu as you did at the beginning of this lesson

You have seen this before. This time all we need to do is click on the name of the file on the right that we want to open. See Figure 1-36 to help explain this.

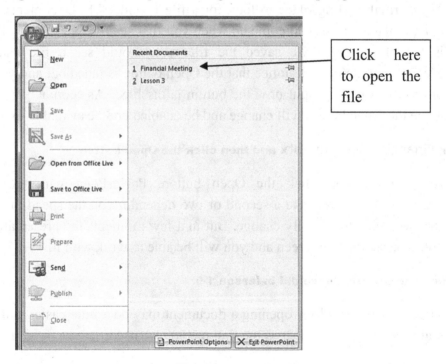

Figure 1-36

Click anywhere outside the Office section to close the menu

Lesson 1 – 11 Printing a Presentation

Let's recap: We can create slides, save the file, close the file, and open a file. Now let's see about printing a presentation. The presentation may be part of a report you are submitting or it may just be your personal presentation. Either way you will probably want a printed copy. Printing the presentation is a simple process, provided that you have a printer set up on your computer.

If it is not open, open the presentation named Financial Meeting

The presentation should be displayed on the screen and look like it did when we last worked on it.

Click the Office Button and move the mouse to the Print choice, but do not click the mouse

There are three choices you can make from this menu: Print, Quick Print, and Print Preview. Before we actually print the presentation, it might be a good idea to see how it will look when we print it. We may find out that the text will not all fit on one page and we may have to do some adjusting to our font size. The available print choices can be seen In Figure 1-37.

Figure 1-37

Click on the Print Preview choice

This will display how the sheet will look when it is printed. There are also a few other options available from this screen (see Figure 1-38).

Figure 1-38

From this screen you can: Close the print preview, view the other pages (if there are any other pages), Zoom in or out on the document, make page settings, or print the document.

In the Page Setup Group we can decide exactly what we want to print. We can print the slides in the presentation or we can print handouts or the note page. The drop down list is shown in Figure 1-39.

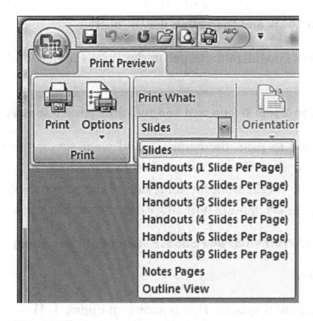

Figure 1-39

Now we will look at the Zoom feature. This feature will allow us increase or decrease the magnification on the Print Preview screen. It will not change the actual presentation.

Click the Zoom button

This will bring the Zoom Dialog box to the screen as shown in Figure 1-40.

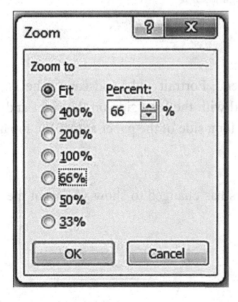

Figure 1-40

There are preset choices that you can zoom to. There is also a percentage box you can use to manually adjust the amount of zoom.

Click the radio button next to 200% and then click the OK button

This will make the print considerably larger in the Print Preview and a lot easier to read.

Click on the 100% button in the Zoom Group and see the difference it makes

Suppose that you decide to print the handouts and want several slides to be on a page? By default the paper will be positioned so that the long side of the paper is going up and down. You might want the long side of the paper to be across the top and bottom. Changing the orientation will allow us to accomplish just that.

Click on the down arrow under Print what and choose Handout (6 per page)

Click on the Orientation button

A short drop down menu will appear. This is shown in Figure 1-41.

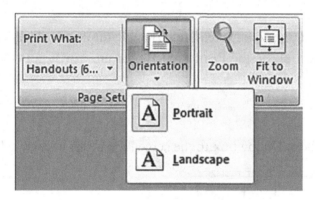

Figure 1-41

There are two choices in this menu, Portrait and Landscape. The Portrait choice is the normal way a paper is printed with the long side on the right and left sides. The Landscape choice prints with the long side of the paper across the top and bottom.

Click on the Landscape choice

As you can see the print previews has changed to show you what the document will look like with this choice.

Change the orientation back to Portrait

The next group we will look at is the Print Group. There are some options we can adjust in this group. These include editing the header and footer sections of the page. We can choose to print in color or print in black and white or with shades of gray. We can also choose the printing order of our slides. This includes should we print the slides in order going across the page or print them in order going down the page. The options drop down menu is shown in Figure 1-42.

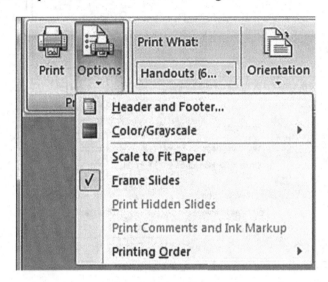

Figure 1-42

The last thing we need to look at in the Print Preview section is the Print button that is in the Print group. Clicking this button will bring the Print Dialog box to the screen.

Click the Print button (the one with the picture of the printer on it)

The Print Dialog box will allow us to choose the printer we wish to use by clicking the downward pointing arrow across from the word Name, assuming, of course, that there is more than one printer configured on your computer.

You can also choose to print all of the slides in the presentation, or the current slide, or you can specify which slides to print. You can also chose to collate the pages (put them in the correct order as they print). If we click OK the document will print. If we click Cancel the Print Dialog box will go away and the document will not print. Figure 1-43 shows the Print dialog box.

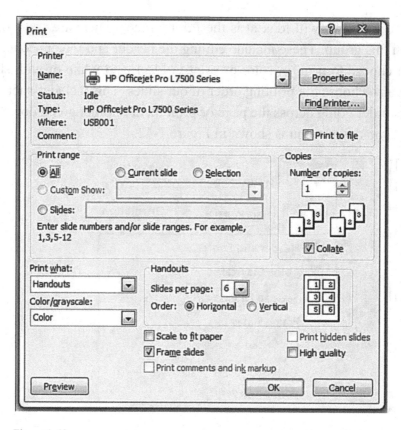

Figure 1-43

Click the Cancel button

Click the Close Print Preview button

Now that we have seen what it is going to look like when we print it, let's see what the other two choices are when we moved the mouse over the print option.

Click on the Office button and move the mouse down to the word Print and then click on the top choice: Print

This will bring the same Print Dialog box to the screen as was shown in Figure 1-43.

In addition to the above mentioned choices we can also choose to print only the slides that we have highlighted. This choice will become available if we actually have some slides highlighted. On the right side we can choose how many copies we want to print.

On the middle left, we can choose to print all of the slides or just the ones we choose. We can also choose to scale the document to fit the size of the paper.

Once we have everything marked correctly, all we have to do is click on the OK button and this will send the information to the printer.

Click Cancel so nothing will print

The last choice is the Quick Print choice and if you click on this choice the entire slide presentation will be sent to the default printer. There will be no dialog box for you to make choices. The slides will just be sent to the printer.

Note: If you click on the print choice from the Office button, the Print Dialog box will come onto the screen. This is identical to the top choice of the three available choices we just discussed.

Chapter One   Review

Microsoft PowerPoint 2007 is a powerful desktop presentation program that will allow you turn your ideas into slide shows and presentations.

The PowerPoint screen consists of: The Office Button, the Quick Access Toolbar, the Title Bar, the Ribbon, the Slides Pane, the Slides and Outline tabs on the left side, and the notes area at the bottom.

The Office Button will allow you to do such things as: open, save, and print a presentation. You can also access the PowerPoint Options from here.

The Ribbon consists of Tabs, Groups, and Commands. If a command is not visible on the Ribbon, it might be found in the Dialog Box that is associated with some of the Groups.

The Quick Access Toolbar will give you quick access to the commands that you decide to put on the toolbar. Commands can be added to the toolbar from either the standard choices or from the More Commands Dialog Box.

Keyboard shortcuts are still available and use the Alt or Ctrl key with a combination of at least one other key. Pressing the Alt key will bring the Key Tip Badges to the screen. From here you can access the Tabs on the Ribbon and the commands on the Quick Access Toolbar. You can also right-click on an object to bring the shortcut menu to the screen.

To create a presentation, you have to decide what type of slide you want in the presentation and then add the necessary data. You will continue this process for each slide. Make sure you save the presentation, every so often, to keep from losing data. The presentation can be closed when you are finished, after it is saved, and re-opened when you need it. The presentation can be printed if you need a hard copy for handouts.

Chapter One   Quiz

1) Pressing the Alt key on the keyboard will bring the Menu Bar to the screen. **True or False**
2) What is shown on the Slides Tab of the PowerPoint screen?
3) In what section of the drop down menu of the Office Button would you be able to look at the properties of the presentation?
4) Where do you tell PowerPoint to check your spelling as you type? Be specific.
5) The Ribbon has three major components. Name them.
6) After you click on the drop down arrow of the Quick Access Toolbar, how do you add an item from the standard choice to the toolbar?
7) Name the Tab and group that contains the Slide Layout command.
8) While running a presentation, name one of the two things that might happen.
9) Name the proper two steps to close a presentation.
10) You can print handouts and have up to nine slides per page. **True or False**

Chapter Two   The Ribbon – A Closer Look

In chapter one, we had an overview of the Ribbon. In this chapter we will take a closer look at the different parts of the Ribbon.

In this chapter we will look at each standard tab of the Ribbon and each group that is on the Tab.

A word of concern: Depending on the size of your monitor, you may not see everything that is shown in the figures. Some of the groups may be condensed and may not show all of the available options at all times. The options are still available, but you may have to click one of the drop-down arrows to see them. I will show you some example of this as we continue.

This chapter will not have much user interface and consists of mostly reading (sorry about that), but it is necessary to have an understanding of the Ribbon.

Note: There are a few other tabs on the Ribbon that are not listed in the standard tabs. These only become available when they can be used. One example is if you insert a picture into your document the Picture Tools will appear with the format tab. These will be discussed when they are available.

Note: As we go through the rest of this book, the various groups and commands will be discussed as we use them.

Lesson 2 – 1   The Home Tab

Open the PowerPoint presentation named Financial Meeting, **if it is not open**

The first thing I will show you is two different views of the same Home tab. One will show the view using a large monitor and the second is with a smaller monitor. You will be able to see the difference between the two and perhaps understand what I was saying on the previous page.

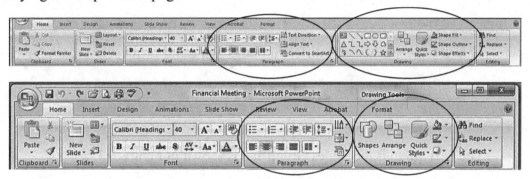

Figure 2-1

If the monitor is smaller the Ribbon will try to compensate by being taller when it is not as wide. Also some of the commands in the different groups may not be visible. Two examples of these are the Paragraph group and the Drawing groups.

As I stated earlier the Home tab contains the commands that one normally uses most often when using PowerPoint 2007.

Let's take a closer look at the different groups on the Home tab.

The first group (on the far left side) is the Clipboard Group and is shown in Figure 2-2.

Figure 2-2

This group deals with the different things you can do with selected text, and objects.

The next group is the Slides Group and is shown in Figure 2-3.

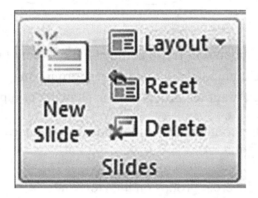

Figure 2-3

This group deals with inserting new slides and the layout of the slides.

The next Group is the Font Group and is shown in Figure 2-4.

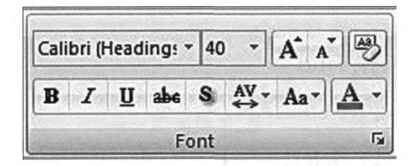

Figure 2-4

As you would expect, this is where you would perform all of the formatting for the text.

The next group is the Paragraph Group and is shown in Figure 2-5.

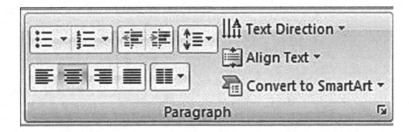

Figure 2-5

This group contains the options used when formatting the paragraphs.

The next group is the Drawing Group and is shown in Figure 2-6.

Figure 2-6

This group contains the objects you can insert into your presentations as well as the styles available for you to use.

The last group on the Home tab is the editing group. This group is shown in Figure 2-7.

Figure 2-7

This group deals with finding and replacing text as well as selecting text and objects.

Lesson 2 – 2   The Insert Tab

Click the mouse on the Insert Tab

The Insert Tab, as you might expect, deals with the various things you can insert into your presentation.

The first group on the Insert Tab is the Tables Group and is shown in Figure 2-8.

Figure 2-8

This deals with inserting tables into your presentation.

The next group is the Illustrations Group and is shown in Figure 2-9.

Figure 2-9

This group deals with inserting pictures and charts into your presentation.

The next group is the Links Group and is shown in Figure 2-10.

Figure 2-10

This group deals with inserting a hyperlink into your presentation and what action is to be taken when you click on or hover over an object.

The next group is the Text Group and is shown in Figure 2-11.

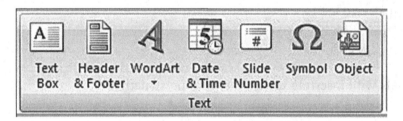

Figure 2-11

This group allows you to insert such things as Text Boxes, Word Art, and Symbols into your presentations.

The last group on the Insert Tab is the Media Clips Group. This group deals with inserting movies and sounds into your presentations. Figure 2-12 shows this group.

Figure 2-12

Lesson 2 – 3   The Design Tab

Click the mouse on the Design Tab

The Design Tab allows you to work with the tools that are used for the way your presentations will look.

The first group is the Page Setup Group. This will let you determine the height and width of the slides as well as the orientation of the pages. Figure 2-13 shows this group.

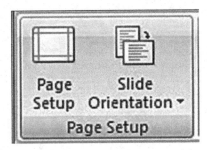

Figure 2-13

The next group is the Themes Group and will allow you to set the general color scheme of all of the slides. Figure 2-14 shows this group.

Figure 2-14

The last group is the Background Group and deals with the graphics and colors that are in the background of the slides. Figure 2-15 shows this group.

Figure 2-15

Lesson 2 – 4 The Animations Tab

Click the mouse on the Animations Tab

The Animations tab will allow you to change how the slides will transition from one slide to the next and if any effects are to be added.

The first group is the Preview Group and will let you preview the slide transition and animations you have on the selected slide. Figure 2-16 shows this group.

Figure 2-16

The next group is the Animations Group which will let you add animation to your slides. Figure 2-17 show the Animation Group.

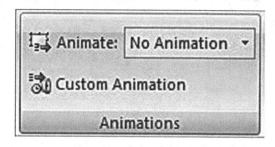

Figure 2-17

The last group is the Transition to this Slide Group. From here you determine how the slides will enter onto the screen. Figure 2-18 shows this group.

Figure 2-18

Lesson 2 – 5   The Slide Show Tab

Click the mouse on the Slide Show Tab

The Slide Show Tab works with how you will present your slide show to others.

The first group in the Slide Show Tab is the Start Slide Show Group. This group deals with how the slide show will start. Figure 2-19 shows this group.

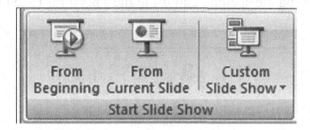

Figure 2-19

The next tab is the Setup group. This group deals with setting up the slide show and recording your narration. Figure 2-20 shows this group.

Figure 2-20

The last group is the Monitors Group. This group will allow you to decide on the screen resolution and things like that. Figure 2-21 shows this group.

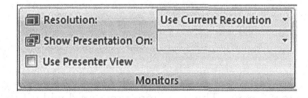

Figure 2-21

Lesson 2 – 6   The Review Tab

Click the mouse on the Review Tab

The Review Tab deals with some of the things you will want to consider before releasing your presentation for others to see.

The first group on the Review Tab is the Proofing Group. This group allows you to do such things as checking the spelling and using the Thesaurus (See Figure 2-22 for this group).

Figure 2-22

The next group on the Review Tab is the Comments Group. This group, as you can imagine, deals with adding comments to your slides. Figure 2-23 shows this group

Figure 2-23

Figure 2-24

The last group is the Protect Group. This is where you can decide if this presentation has restricted access or the access to the presentation is unrestricted.

Lesson 2 – 7   The View Tab

Click the mouse on the View Tab

The View Tab will allow you to view and work with your slides in different ways.

The first group on the View Tab is the Presentation Views Group. This group will let you decide which layout you want to view on the screen as well as editing the slide layout. Figure 2-25 shows this group.

Figure 2-25

The next group is the Show/Hide Group. This will allow you to show such things as the ruler and gridlines to help make sure everything is positioned correctly on the slides. Figure 2-26 shows this group.

Figure 2-26

The next group is the Zoom Group. This will let you see the slides in larger and smaller views when you are working on them. Figure 2-27 shows this group.

Figure 2-27

The next group is the Color / Grayscale Group. This will allow you to view the presentation in color, black and white, or with various shades of gray. Figure 2-28 shows this group.

Figure 2-28

The next group is the Window Group. This group deals with viewing different screens when you have more than one window on the screen. Figure 2-29 shows this group.

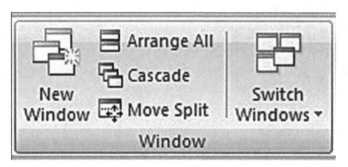

Figure 2-29

The last group on the View Tab is the Macro Group. This group will allow you to create and run macros which under the right circumstances can save you a lot of time. Figure 2-30 shows this group.

Figure 2-30

Lesson 2 – 8   The Developer Tab

Click the mouse on the Developer Tab if it is visible

The Developer Tab will allow you to add controls such as buttons and checkboxes to your slides. The only thing new about this is that you will also have to add the code (programming) to make these controls work. Teaching you how to write Visual Basic code is well beyond the scope of this book. I will show you the Developer Tab, but we will not be covering it in this book.

The first group on the Developer Tab is the Code Group. This will bring up the Visual Basic programming part of PowerPoint to the screen where you can write the software. Figure 2-31 shows this group.

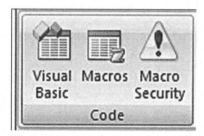

Figure 2-31

The other group is the Controls Group. This group will allow you add the controls to the slides. For the controls to actually do anything you will need to add the programming code. Figure 2-32 shows this group.

Figure 2-32

Note: There are a few other tabs that are not part of the main tabs. These tabs will only appear when they can be used. They will be covered as we use them.

Chapter Two   Review

The Home tab contains the commands that one normally uses most often when using PowerPoint 2007.

The Insert Tab deals with the various things you can insert into your presentation.

The Design Tab allows you to work with the tools that are used for the way your presentations will look.

The Animations tab will allow you to change how the slides will transition from one slide to the next and if any effects are to be added.

The Slide Show Tab works with how you will present your slide show to others.

The Review Tab deals with some of the things you will want to consider before releasing your presentation for others to see.

The View Tab will allow you to view and work with your slides in different ways.

The Developer Tab will allow you to add controls such as buttons and checkboxes to your slides.

Chapter Two   Quiz

Name the Tab and Group where you will find the following commands.

1) New Slide
2) Insert a chart
3) Transition sounds
4) Align text
5) Run a Macro
6) Insert a movie clip
7) Find text
8) Add a comment
9) Slide orientation
10) Record a narration for the slide show

Chapter Three   Editing a Presentation

Almost without exception, you will create a slide show and find a better way of communicating your thoughts. It may turn out that you need to move text to a slide in the presentation or you may need to add or delete certain text. This chapter will deal with the various things available when you are editing your presentation.

Lesson 3 – 1   Inserting Slides

Open the PowerPoint presentation called Financial Meeting

This file is located in the My Documents folder (if you are using Vista the folder is called Documents).

This presentation has two slides in it and for this lesson we will add another slide. The first slide has the title for our presentation and the second slide goes right into the income part of the meeting.

Click on the New Slide icon in the Slides Group of the Home Tab of the Ribbon

Wow, that was a mouthful! Your screen should look like Figure 3-1.

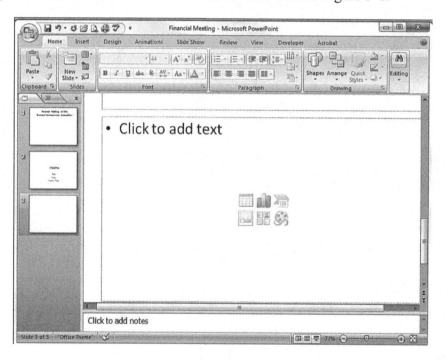

Figure 3-1

If we were going to insert a table or chart or even a picture into our presentation, this would be the slide layout we would need. The only problem is we are going to add more text to our slide. We will need to change the layout of the slide.

Click on the Layout command and choose Title only

This is the same slide layout we started with in slide one.

Using the mouse click inside the box that says Click to add title

Add the following text (be sure to press Enter after the first two)

Income

Expenses

Profit / Loss

Your slide should now look like Figure 3-2.

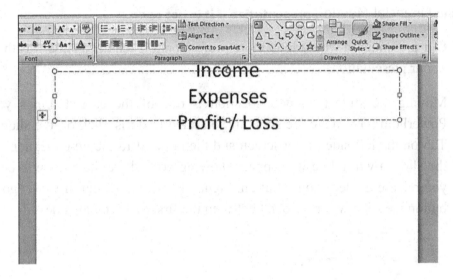

Figure 3-2

I think you would agree that this is not a very pleasing slide to gaze upon. We will make it look a lot better, but that is another lesson.

Save your changes

Lesson 3 – 2 Moving Slides

In lesson 3-1 we learned how to insert a new slide into our presentation. The slide we inserted is not in the correct place. If we ran the presentation like it is it would look really silly. We need to change the order of the slide so they will make sense.

Open the Financial Meeting **presentation if it is not open**

Our presentation would look more professional if the third slide was the second slide in our presentation.

Moving the slide to a new location is one of the easiest things you can do in PowerPoint. To move the slide all you have to do is click on the slide in the Slides Tab on the left side of the screen and then drag it to the new location. As you move the slide, a white line will appear showing you where the slide will be positioned if you release the left mouse button. Figure 3-3 shows us that if we let go of the mouse button the slide will be moved between the first and second slide.

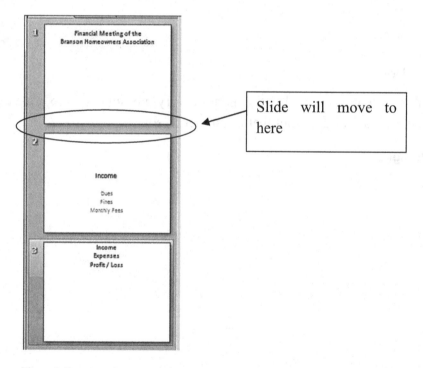

Figure 3-3

Using the mouse, click on the third slide and hold the left mouse button down and drag the slide up until it is between slides one and two and then release the left mouse button

Save your changes

Lesson 3 – 3   Editing Slides

If necessary open the Financial Meeting **presentation**

In this lesson we will do one simple task. We will clean up some of the slides and the text will look like it belongs in the slides.

Click on the first slide in the presentation

This is the first slide we will adjust. In this slide the text is at the top of the slide and it should be in the center of the slide. There is more than one way to move the text into the center of the slide.

First we will look at the way you will probably use the most.

Click somewhere inside the text on the slide

When we click inside the textbox an outline will appear around the words. This dotted outline shows the current boundaries of the textbox. We a can adjust these boundaries to allow the text to flow toward the center of the slide. You will notice that there are small circles and squares at the corners and in the center of the outline around the textbox. These are called sizing handles and they are what we use to change the size of the textbox. To use these sizing handles you simply click on one of them and drag the boundary to a new size. Figure 3-4 shows the sizing handles.

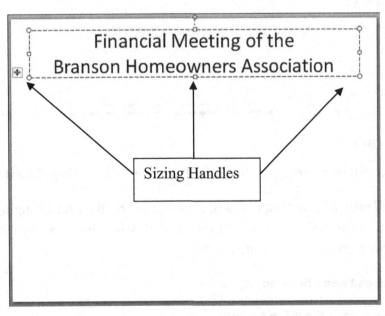

Figure 3-4

Move the mouse pointer until it is on the lower center sizing handle

You will know that the mouse is in the correct place because it will change into a small black double sided arrow. In this case the arrow will be pointing up and down. This will allow you to change the height of the box. If you move the mouse pointer to either one of the handles on the side, the arrow would point to the right and left and you could change the width if the textbox. If you moved to one of the handles that are in the corners, the arrows would be at an angle and you could change both the height and width of the textbox.

Press and hold the left mouse button down and drag the mouse downward until the textbox covers the slide

Your slide should now look like Figure 3-5.

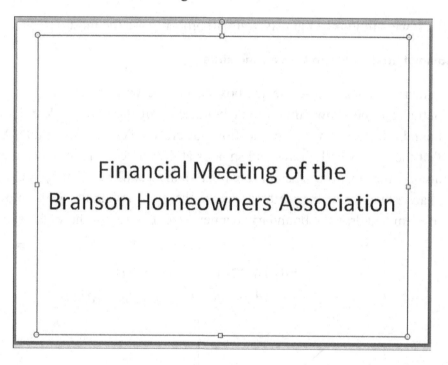

Figure 3-5

The text will be centered on the slide. The slide will now look like it should look.

You will probably be tempted to use the second method for changing the position of the text on the slide. I am not saying that this is a bad way to move the text, but somehow it doesn't seem quite right.

Click on the second slide to select it

Click the mouse before the "I" in Income

Press the Enter key six times

This will move the text down six lines and it should be about in the center of the slide. There is nothing wrong with doing it this way and it will work, however anyone that is allowed to edit the presentation will notice that you didn't expand the textbox. Functionally it will still work.

Save your work

Lesson 3 – 4 Adding Text

Now that you can move the text around inside the slide we will look at how to change the text that is on the slide. Changing the text is about as easy as changing the position of the text on the slide, perhaps even easier.

If necessary open the Financial Meeting **presentation**

Click the mouse on slide 1 to select it

Click the mouse before the first letter of the text (before the F)

Now that we have our insertion point (the flashing vertical line) where we need it we can add some text to the slide.

Type the following text

> Welcome to the

Press the Enter key on the keyboard

Your text should now look like Figure 3-6.

Figure 3-6

That is all there is to adding text to a slide. Just to make sure you have it let's add some text to slide two.

Click on slide 2 to bring it to the screen and then we can edit it

Using the same technique as we did for slide one add the following text to the beginning of slide2

Monthly Meeting Agenda

Press the Enter key after you type Agenda

Just a reminder: You could have entered the text anywhere in the slide by clicking the mouse where ever you wanted to insert the text. It doesn't have to be at the beginning. Your second slide should look like Figure 3-7.

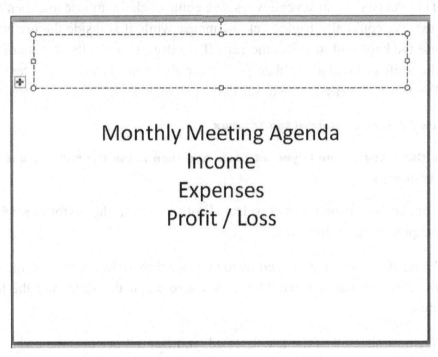

Figure 3-7

Save your work

Lesson 3 – 5 Selecting, Replacing, and Deleting Text

Open the file Financial Meeting **if it is not open**

In the last lesson we learned how to add text to a slide. In this lesson we will see how to select text and replace or delete the text.

Click on slide 1 if it is not the current slide on the screen

Select all of the text in the textbox

This can be done in several ways. We could click the mouse and then drag it over the text. We could also triple-click the mouse while it is inside the textbox, or we could use the keyboard to select the text. If we decide to use the keyboard we could hold the Shift key down and then push the right arrow key until the text is highlighted. But there is an easier way to select the complete text.

First click the mouse inside of the Textbox

Hold the CRTL key on the keyboard down and then press the "A" key one time and then release both keys

This keyboard shortcut will select all of the text in the textbox. Keyboard shortcuts are pretty cool, aren't they?

Note: If you had not clicked inside the Textbox before you used the shortcut, using the Ctrl + A shortcut would have selected all of the slides not the text in the first slide.

We now have all of the text selected, you can tell because the text is highlighted. If we wanted to make a change to the text or even delete it now would be the time to get this done, while the text is highlighted.

The easiest way to replace or delete text is first to select the text. You can use the mouse to select the text by moving the mouse pointer to the desired location and pressing and holding the left mouse button down and then drag the mouse over the text you want to select. This is usually the preferred method if you are selecting more than one word. If you are only selecting one word the easiest way is to double-click on the word you want to select.

Click on slide 2

Double-click on the word "Monthly" in slide two

Once you have the word selected, you can replace it by simply typing in the new word(s). The highlighted word will go away and the newly typed word will replace it.

Type the word

> Quarterly

Note: You may need to add a space between the first two words to separate them.

What if we decide to use this same presentation for our meeting at the end of the year?

Replace the word "Quarterly" with the word "Annual"

As you can see, replacing existing text is a simple process. Now here are a few words on deleting text.

To delete text you could click either just before or just after the text you wanted to delete. If you clicked just before the text, you would need to press the "Delete" key on the keyboard until the text is removed. If you clicked just after the text, you would press the "Backspace" key on the keyboard until the text is removed.

Remember every time you press the Delete key one letter to the right of the insertion point will be deleted. In like fashion, every time you press the backspace key one letter to the left of the insertion point will be deleted.

You could also click and drag the mouse over the text you want deleted and then press the delete key. This would also delete the selected text.

On slide two delete the word "Annual"

Save your work

Lesson 3 – 6   Using Cut, Copy, and Paste

This lesson is dedicated to moving text, and objects, around in your presentation. We will be using Cut, Copy, and Paste to accomplish this. First, let me give you a brief description of these commands.

Cut: This will <u>remove</u> any data that is highlighted (selected) and move it to the clipboard. The clipboard is a temporary storage place and will temporarily hold the data until you can put it someplace else in the presentation or even into a completely different presentation or program.

Copy: The copy command will <u>make a copy of the selected text</u> (or object) and place it in the clipboard for you to use later.

Paste: The paste command will <u>copy from the clipboard</u> and put the information into the presentation.

Once you have selected the text, you can move it to another place in the presentation by cutting and then pasting it elsewhere. Cutting and pasting text is one of the most common things you will do when editing a presentation. When you cut text, it is removed from its original location and put it in a temporary storage area called the Clipboard. You can then move the insertion point to a new location and paste the text from the Clipboard. The Clipboard is available from any program in Windows, so you can cut text from one program and paste it into another program.

Before we start cutting, copying, and pasting let's add a new slide to our presentation.

Open the file Financial Meeting **if it is not open**

Add a new slide and make it a title only slide

If you need a short refresher course go back to lesson 3-1.

Select the text Dues, Fines, and Monthly Fees from slide 3

When we have the text selected we can use the Copy button on the Ribbon to make a copy of the text and place it on the Clipboard. The Copy command is in the Clipboard Group which is on the Home Tab of the Ribbon (See Figure 3-8).

Figure 3-8

Click on the Copy command

This will place the selected text on the clipboard.

Click inside the textbox where the title goes and then click on the Paste command

Your slide will probably look like Figure 3-9.

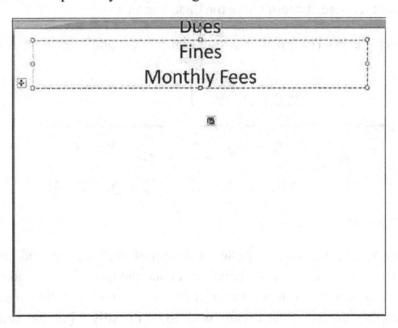

Figure 3-9

This doesn't look so great all of a sudden. Let's see if we can fix our little problem.

Using the sizing handles pull the slide's textbox down toward the bottom of the screen until the text is about in the middle of the slide

Now we are looking a lot better.

The only difference between copying text and cutting text is that the text will disappear from the original location if we cut it. The text will stay in its original location if we copy it. The paste command works the same for both cut and copy.

Save your changes

Lesson 3 – 7   Using Undo and Redo

Before we continue on in our discussion of Editing, there is one other thing that will come into play: Using the Undo and Redo feature.

Being human, sometimes we make mistakes. If I had not been forced to use this feature over and over again during the writing of this book, it probably would not seem as important as it actually is. Microsoft, in their infinite wisdom, looked into the future and knew that I was going to use their product and added this feature probably just for me. I will explain it to you just in the off chance you may need it.

Guess what? Undo is not on the Ribbon. Just when you thought the Ribbon held everything you would ever need, it doesn't have the Undo button.

The Quick Access Toolbar is shown in Figure 3-10, and is the home of the Undo and Redo buttons.

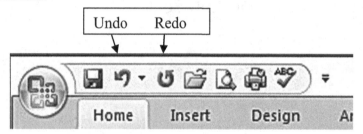

Figure 3-10

Any time you copy, type, cut, paste, or do almost anything the Undo button becomes available. If you make a mistake, you can click the Undo button and everything will be as it was before you made the mistake. Now let's be realistic here, PowerPoint will not know that you didn't really want to do the silly thing that you just did. That means that you can't make the mistake today and tomorrow when you realize that you made the mistake, expect PowerPoint to undo it. If you click the Undo button, PowerPoint will undo the last thing that you did, not the mistake you made five minutes ago. Obviously this may not be magic, but it is close.

If necessary open the Financial Meeting **presentation**

Click on the first slide

Press the Delete key on the keyboard

Oh No! That wasn't the slide I wanted to delete. I wanted to delete the forth slide not the first. What can I do?

Before you panic, we can have the Undo feature fix our problem.

Click the Undo button on the Quick Access Toolbar

The first slide comes back and is now safe and sound in our presentation. Wow, that was a life saver.

After you use the Undo button, the Redo button will be available for you to use. Before you used the Undo button, the Redo button was faded out and not available. Now it is available.

If you decide that you really did want to delete the first slide, you can use the Redo button to redo what you just undid. Basically the Redo button will undo the undo.

Lesson 3 – 8   Using Spell Check

There is probably nothing more embarrassing than to have someone come up to you and correct your spelling in a presentation. Microsoft thought that this would be embarrassing for you as well and provided you with a spell check feature. Before you let anyone see your work, it is probably a good idea to run the spell checker.

Did you ever notice that when you are typing a word and press the space bar, a red line appears under the word? This is because PowerPoint automatically checks the spelling of words as you type them. Many people think this is the greatest feature of PowerPoint: the Spell Checker. PowerPoint not only checks for spelling errors while you type, but it also highlights the spelling errors with a red underline.

PowerPoint can also correct common mistakes for you as you type. If you typed hte PowerPoint could automatically change it to the.

Open the file Financial Meeting2.pptx

This file can be found with the files that were downloaded from the website. If you did not download the files from the website, go back to the front of the book and read the page with Important Notice at the top. The files will probably be in the PowerPoint 2007 Lessons folder under the My Documents folder.

This presentation has two misspelled words in it. We will use the spell check feature to find and correct these mistakes.

You will immediately notice that one of the words has a red underline and is obviously misspelled.

The Spell Check command is in the Proofing group which is on the Review Tab of the Ribbon (See Figure 3-11).

Figure 3-11

Click the Spell Check Icon

Spell Check will start searching through the presentation to find any misspelled words. It will stop searching when it comes to the word "Finanical". Spell check will now bring the Spelling and Grammar Dialog box to the screen as shown in Figure 3-12.

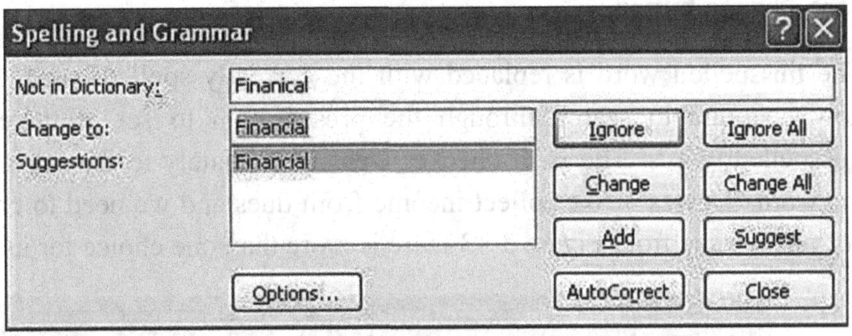

Figure 3-12

From this dialog box we can learn a few things. First we can see that the word in question is not in the dictionary. We are also given suggestions for words that we can use in place of the unknown word.

On the right side are more choices for us to make. We can choose to ignore this occurrence of the word or all occurrences of the word. That brings up an obvious question is why would you want to ignore a misspelled word? One possibility is that some companies intentionally misspell a word as part of their logo or product name. As examples; some trucking companies have "Xtra" and "Xpress" on the side of their trucks.

If the word is correctly spelled, and you are sure that it is correct, you can add the word to the dictionary. I have added my e-mail user name to the dictionary so that it will not show up as a misspelled word.

If the correct spelling is shown under the suggestions, you can highlight the correct spelling and then click on change to replace your misspelled word with the correctly spelled word. If you wish you can replace all occurrences of this word (the one that is not in the dictionary) with the correctly spelled word.

You can also choose the auto correct option. This will allow PowerPoint to decide what to replace the misspelled word with. PowerPoint will replace the misspelled word with what it thinks is the best possible choice. Be careful with this option, it may not be your best choice.

If there is more than one possible choice for the correctly spelled word you can click the suggestions button to move the focus from the change to box to the suggestions list.

If you want, you can also cancel the spell check by choosing the Close choice.

Our word is not spelled correctly and we need to change it to the correctly spelled word "Financial".

Click on the Change button

The misspelled word is replaced with the correctly spelled word. PowerPoint will now continue to search through the presentation to see if there are any other misspelled words. The spell checker went immediately to the third slide and found the word "Dewes". We collect income from dues and we need to change this word. As you can see from Figure 3-13 there is more than one choice for us to pick from.

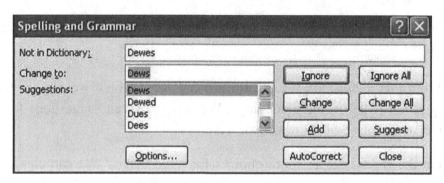

Figure 3-13

If we were going to use the AutoCorrect feature PowerPoint would substitute the misspelled word with the word Dews. This is not the correct word for this situation. We meant to put Dues in the presentation.

Click on the word Dues under suggestions and then click the change button

When you click the change button PowerPoint will continue searching for misspelled words and jump to the last slide. I want to show you something different for the last slide.

Click the Close button

The Spell Check feature is now closed and we can look at an alternate way to replace a misspelled word.

Click on slide four to bring it to the screen

You can still tell that the word Fewes is misspelled because it has a red underline. This is shown in Figure 3-14.

84

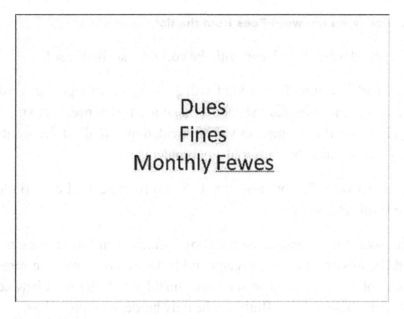

Figure 3-14

Move your mouse pointer until it is on the misspelled word and then right-click the mouse

This will bring a shortcut menu to the screen. If you right-clicking on an object it will bring a shortcut menu to the screen showing you all of the things you can do with this object. Figure 3-15 shows the shortcut menu.

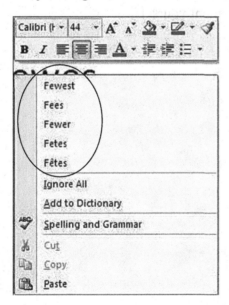

Figure 3-15

There are two sections to the menu. The top part has the formatting commands that are available (these will be discussed later), and the bottom section has editing commands that are available. The top half of the lower section has suggestions for the correct spelling of the misspelled word.

Using the mouse click on the word Fees from the list

The misspelled word is replaced with the correctly spelled word.

One word of caution: PowerPoint will pick up on misspelled words; however it won't pick up on sentences that don't make a lot of sense. As an example, if you were typing a short paragraph in your presentation and all of the words were spelled correctly, PowerPoint may not find any problems.

The road between Saigon and the U.S. army base at Long Binh was heavily burdened will sellers.

I ran the Spell Check feature on the above sentence and there were no errors found. All of the words are spelled correctly and there are no grammar errors. The sentence still does not make sense. The sentence should say; "The road between Saigon and the U.S. army base at Long Binh was heavily burdened with sellers."

I also ran the spell check on the words below in a different slide and no problems were found.

In this slide we don't have no words that make sense.

The moral of this story is that the Spell Check is a great tool but it will not replace you actually proof reading the document yourself.

Save your changes in My Documents

Close the presentation

Lesson 3 – 9   Finding Information

Searching for text in your presentation can be time consuming and frustrating. You know that somewhere in your presentation you referenced a certain word, but now you are trying to find it. Oh Boy! How much fun is this?

You could search through the entire presentation, word after word and slide after slide, or you could let PowerPoint do the search for you.

For this lesson we will use a presentation that has several slides and it is a little more difficult to find one specific word.

Open the file Lesson3.pptx

This file is with the files that you downloaded. There are 47 slides in this presentation and finding a particular word will be a time consuming project.

On the Home tab there is a group which deals with finding text in your presentation. This group is the Editing Group and is shown in Figure 3-16.

Figure 3-16

Click the Find button in the Editing Group

The Find Dialog box will come to the screen (See Figure 3-17).

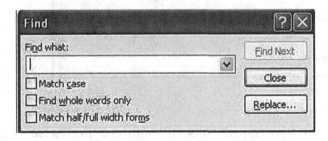

Figure 3-17

Under the Find what section you can type any word that is in the presentation and PowerPoint will search and find it. This will only work for text that has been typed in the textboxes. If the word you are looking for appears as part of a picture that was inserted into your presentation it will not be found. I know that what I just said sounds silly, but as you can imagine it has come up before.

Somewhere in this presentation I discussed using the right mouse button, but I can't remember where.

In the Find what box type Right **and then click the Find Next button**

The Find command should take you to slide 20 and highlight the word Right.

Click the Find Next button to see if there is another occurrence of the word right

PowerPoint will start searching again and if there is another occurrence it will be highlighted.

As you continue to click the Find next button PowerPoint will stop at every occurrence of the word right and there are several.

What if we decide that one of the words we used in the presentation is not actually the best word for what we were trying to explain? We can use the Replace command to find a word and replace it with a different word.

Click on slide 1 to bring it back to the screen

Click on the Replace command

The Replace Dialog box will come onto the screen (see Figure 3-18).

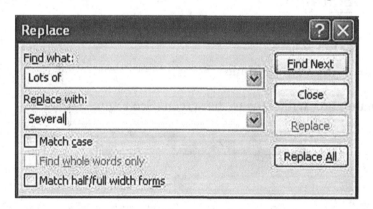

Figure 3-18

In the figure above I have typed "lots of" into the find what box and I typed "several" in the replace with box.

Type lots of in the Find what box and several in the Replace with box

Click on the Find next button

PowerPoint will start searching for the requested words and stop if it finds them. In the above example we typed in two words and PowerPoint will only stop searching if it finds both words together. It will not stop on the word "lots" or the word "of", it will only find a match if it finds the words "lots of" together.

When PowerPoint finds the requested words the Replace button will change color and be available for us to click. We have a choice to make. We can replace the words or we can replace all occurrences of the selected words.

A word of caution: If you click on the "Replace All" button PowerPoint will replace every occurrence that it finds. It will not ask you if you are sure, it will just carry out the command. Be very careful with this, it may not be what you want to do.

Click on the Replace button

PowerPoint will replace the words "lots of" with the word "several" and start searching again to see if there are any other occurrences of the words "lots of". There are no other occurrences so PowerPoint will bring an information box to the screen to let you know that it is finished searching (See Figure 3-19).

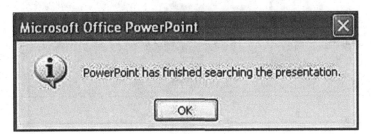

Figure 3-19

Click the OK button

There is another choice under the Replace command. This is the replace font command. This will bring a dialog box to the screen which will allow you to replace one font type with another font type. It works just like the replace text dialog box.

Save your work in My Documents

Close the presentation

Lesson 3 – 10   Adding Comments

The last thing we are going to discuss in the editing chapter is adding comments to your presentations.

Adding a comment to a presentation is quite easy and can be done in only a few moments. I guess the big question is why would you add a comment? I suppose that you could use the comment to remind yourself of something, perhaps something that you need to say for a presentation.

Open the Financial Meeting **presentation**

In this presentation we outlined the steps for our meeting. We may also be adding more slides as we continue on in the book. Now we will add comments that we will refer to as we go through the meeting.

It might surprise you to know that when you insert comments you do not use the Insert Tab. You add comments from the Review Tab. The Comments Group of the Review Tab is shown in Figure 3-20.

Figure 3-20

Click inside the textbox of slide 1 and move the insertion point to the very end of the last line

You can move the insertion point by clicking at the very end of the last line with the mouse or you can use the arrows on the keyboard to move the insertion point.

Click the New Comment command

This will open the Markup area of the screen. This is a part of the screen that is usually not visible and is used to show you the comments that are in the presentation. Figure 3-21 shows the markup area.

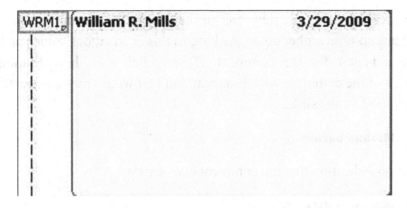

Figure 3-21

On the right is the actual comment area. First we have my initials and the number one. My initials are there because I am the author and the number one because it is the first comment. My full name is also shown as it is in the PowerPoint setup screen. The date the comment was made is also shown. The insertion point is already in the comments box and is ready for you to start typing.

Type the following

The people at this meeting usually don't want to be here, so make some small talk to relax them.

The results should look like Figure 3-22

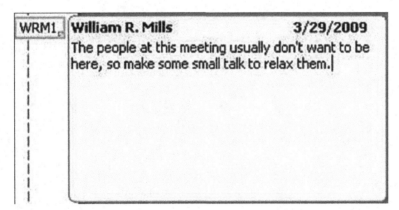

Figure 3-22

Click anywhere outside of the comment section

The actual comment area will disappear and only the small yellow square with your initials will be visible.

Notice back up on the Review Tab in the Comments Group that the command for Show Markup is an amber color. As long as this is an amber color you will be able to see the indicator for the comment. If you click the Show Markup button, the indication of the comment will disappear and you would never know that there was a comment tied to this slide.

Click the Show Markup button

There is no indication that the comment even exists.

Click the Show Markup button again

Now you can see the comment indicator.

Click on the indicator to see the actual comment

Click outside of the comment to only see the indicator again

If you need to edit a comment, all you have to do is click on the Edit Comment command on the Ribbon.

Save your work and close the file

Close PowerPoint

Chapter Three   Review

Editing a presentation will probably be a large part of preparing your presentations.

New slides are inserted into your presentation by clicking on the New Slide command. You can also insert a new slide by using the keyboard shortcut Ctrl + M. If you want to add a slide after a specific slide, you would click on the slide that will be just <u>before</u> the new slide and then use one of the previously mentioned techniques for inserting a slide.

You can move a slide to a new location by clicking on it and dragging it to the new location.

Textboxes inside the slides can be resized by dragging the borders using the sizing handles.

Text can be added to a cell by clicking inside the textbox at the desired location and typing the new text.

The easiest way to replace or delete text is to first select the text and either type over the text or delete it.

To move text from one location to another, use cut, copy, and paste.

You can use the Undo command to undo the last action you performed, and use the Redo command to put it back.

Remember to spell check all presentations.

The Find command will let you search for a word or phrase. The Replace command will let you replace a specific word or phrase with another.

You can add comment for your use, usually to remind you of something pertaining to the slide.

Chapter Three   Quiz

1) List the steps required to insert a new slide between slides 2 and 3.
2) If you are dragging a slide to a new location, what indicator will there be to let you know where the slide will move to if you release the left mouse button.
3) If you increase the height of a textbox, the text inside the box will also move in the same direction as the height is changing. **True or False**
4) If you want to add additional text inside a textbox, just click the mouse where you want the text and start typing. **True or False**
5) If you double-click the mouse inside a textbox, all of the text inside the textbox will be selected. **True or False**
6) The Paste command will move the selected text from one location in the presentation to another place in the presentation. **True or False**
7) The Undo command will search the presentation and undo any mistakes it finds. **True or False**
8) On what tab and in which group is the Spell Check command?
9) Using the Replace Dialog Box you can search for a word and replace all occurrences of it with another word by clicking the Replace All button. **True or False**
10) Comment, once in place are always visible. **True or False**

Chapter Four   Managing Files

This chapter is not really a part of Microsoft Office 2007. It is included in this book because it is so important. After you make your presentations you will need to save them in a place that will enable you to find them easily when you need them.

If you are one of those people who have your entire monitor screen filled with icons and your documents folder is so full you have trouble finding the correct file, this chapter may change your computer life.

In this chapter we will learn how to make folders and how to move your presentations around so that each presentation gets in the correct folder. You will learn how to organize your computer.

Just so that you remember, these presentations we have been making are actually files that you are storing on your hard drive.

Lesson 4 – 1   Making New Folders

If you were at home storing papers you probably wouldn't put them in a big pile in the center of your table or desk. You would more than likely put them in a filing cabinet. In your filing cabinet you would not just randomly push papers into any drawer. You would have several folders to keep similar documents together, and have the papers all neatly stacked inside each folder. This is the same line of thought you need to have with your computer.

Before you can organize your files, you need a place to keep them. Our first lesson is dedicated to teaching you how to make folders.

Click on the Start button and select My Documents **on the right side**

The Windows Explorer program will jump onto the screen. On your screen you will have one of two possible views. The left side of the screen may have "My Documents" at the top and it will be highlighted (see Figure 4-1), or the left side may not list any of the folders at all (See Figure 4-2).

Figure 4-1

Figure 4-2

The right side of the screen will show all of the files and folders that are inside the folder. The My Documents folder is an ideal place to save your files. However putting all of your files in the My Documents folder can also make this folder cluttered and make it hard to find things.

If you had other folders inside of the My Documents folder you could start to organize your computer.

Let's make another folder inside of the My Documents folder. If you have the view that is shown in figure 4-2 making a new folder is quite easy, you click on Make a new folder. If you have the view that is shown in figure 4-1 it is a little harder. Just to have everyone on the same page, if you have the view that is shown in figure 4-2 click on the folders icon that is shown in Figure 4-3.

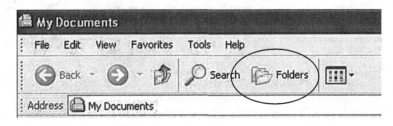

Figure 4-3

We all have the same view and can work together.

Click on File **in the menu bar at the top**

Move the mouse to New **and then click on** Folder

Figure 4-4 shows this process.

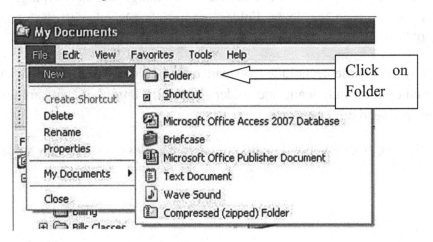

Figure 4-4

97

If you check your computer screen you will find a new folder underneath the My Documents folder. It is named New Folder (See Figure 4-5).

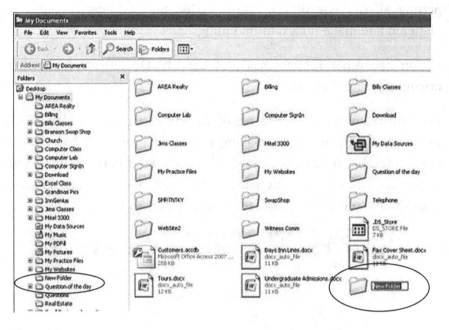

Figure 4-5

The actual folder on the right has the name highlighted and you are expected to give this folder a different name. If you are going to use this folder to store files it should have a name that reflects the files that are stored in it. Let's use this folder to store the files that we will make as we travel through this book.

Using the keyboard type PowerPoint 2007 **and then press the Enter key**

You do not have to click the mouse or anything, just start typing. You will notice that as you type the "New Folder" name is replaced with the name "PowerPoint 2007".

Using this same technique you can create folders to keep all of your like files together. You may want one folder for family letters or that famous Christmas let that you send out every year.

In the next lesson you will learn how to move files into your new folder.

Lesson 4 – 2   Moving Files

Let's put the Presentations we have been using inside the PowerPoint 2007 folder. In this lesson I will show you how to move a file (presentation) from one folder to another folder.

I asked you way back in lesson 1-8 to make sure you saved all of your presentations in the My Documents folder. Now we are going to move them to the PowerPoint 2007 folder. If you didn't save them in the My Documents folder, I don't know where they will be, you are going to have to look for them. To find them you may have to click on Start and then search and then all files and folders, then you can enter the name of the presentation (such as Financial Meeting) and then click search. In the results section, you will be able to see where the file is located.

The rest of this lesson is going to assume that your presentation is located in the My Documents folder.

Click on the Start button and then click on My Documents.

You should see, somewhere in there, a file that has the name Financial Meeting.pptx

Click on the file named Financial Meeting.pptx **and drag it over to the folder named PowerPoint 2007 and then release the mouse button**

As you pass over the folders, each one will highlight to let you know that if you release the mouse button this is the folder your file will be moved to. When the folder named PowerPoint 2007 is highlighted, you can release the left mouse button. The file named Financial Meeting should now be gone from the My Documents folder. Let's see if it in the PowerPoint 2007 Folder.

Double-click on the PowerPoint 2007 folder to open it

The results should look something like Figure 4-6.

Figure 4-6

That is all there is to moving files from one folder to another.

Armed with this knowledge, you can start making folders to keep similar documents together. Go ahead, start making folders and move your files. Now you can start organizing your computer.

Lesson 4 – 3   Copying Files

Just for the sake of having something to do, we want to keep our presentation in the PowerPoint 2007 folder, but we also want the Financial Meeting presentation file to be in the My Documents folder. To accomplish this we will need to copy the file from one folder to another, not move it.

Double-click on the PowerPoint 2007 folder to open it if it is not already open

Right-click on the file Financial Meeting.pptx **and select** copy **form the menu** (see Figure 4-7)

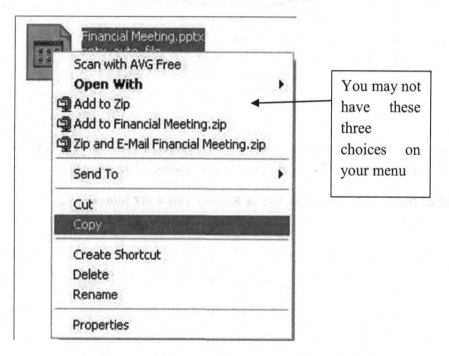

Figure 4-7

101

Right-click on the folder My Documents **and choose** Paste **from the menu** (see Figure 4-8)

Figure 4-8

Now the file "Financial Meeting" is located in both folders (PowerPoint 2007 and My Documents). That is all there is to copying a file.

Copy all of the downloaded files to the PowerPoint 2007 folder.

Chapter Four   Review

To organize your computer, make folders and keep similar documents together.

You can move a file to another folder by dragging it to the new folder. You can also right-click on a file and choose the cut command.

You can right-click on a file and choose the copy command if you want to have the file in more than one folder.

Chapter Four   Quiz

Make a new folder under the My Documents folder called Test and put a copy of the Financial Meeting presentation in the folder.

Chapter Five   Formatting

You have probably seen presentations that use different fonts, italicized and boldfaced type, and fancy paragraph formatting. This chapter will explain how to format both characters and paragraphs. You will learn how to change the appearance, size, and color of the characters in your presentations. You will also learn how to format paragraphs: aligning text to the left, right, and center of the slide. You will also learn how to create bulleted and number lists.

Knowing how to format characters and paragraphs gives your presentations more impact and makes them easier to read.

Lesson 5 – 1   Formatting Text using the Ribbon

PowerPoint allows you to put emphasis on text in your presentations by making the text darker and slightly heavier. This is called Bold. You can also make the text slanted (italics), or make the text larger (or smaller), or you can use a different typeface.

The easiest way to apply character changes is to use the Font Group of the Home tab of the Ribbon.

Open the Financial Meeting **presentation**

Click on the first slide

We will use the Font Group to make some simple formatting changes. This group is found on the Home tab of the Ribbon and is shown in Figure 5-1.

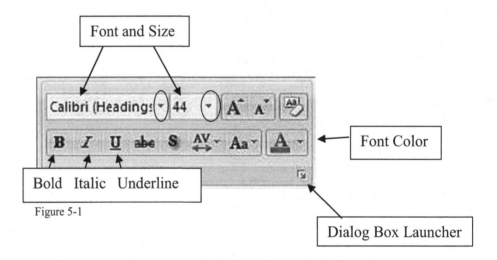

Figure 5-1

Using the mouse, select the first line of the slide (Meeting Agenda)

We are going to change the font and size of the text.

Click the down arrow after Calibri (Headings)

This will bring the drop down list for the fonts to the screen. This will list all of the available fonts, and we can choose a new font from the list.

Scroll down the list and click on Castellar (See Figure 5-2)

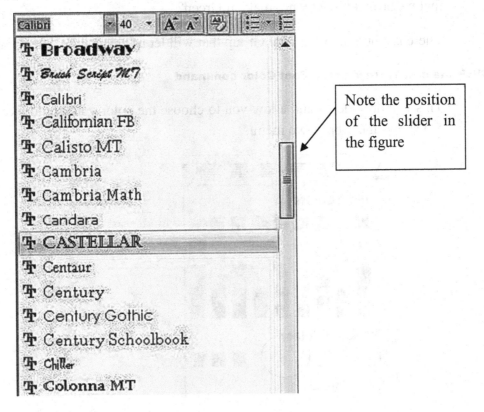

Figure 5-2

Next we will want to change the size of the text and make it larger.

Click the down arrow next to the font size (40) and choose the size 48

The first line now has a different font than the rest of the text and is physically larger. The first line will now stand out from all of the other text. That was fun, let's try some other things.

Select the second line (Income)

The size of this text seems to work fine, so let's only change the way it looks.

Click the mouse on the "Bold" command

Click the mouse on the "Italic" command

Click the mouse on the "Underline" command

Now the text stands out but it could use one more thing. Since this represents money that we brought in why not make it Green?

There is a button in the Font Group that will let us change the color of the font.

Click the down arrow on the Font Color command

This will drop down and allow you to choose the color you would like to use. Figure 5-3 shows the drop down menu.

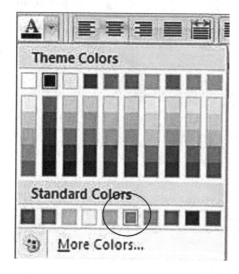

Figure 5-3

Click on the darker green under Standard Colors

Want to play some more?

Click the down arrow on the Font Color command

If you do not see a color that you like, you can always click on the More Colors option button at the bottom.

Click on the More Colors button

New choices will appear for you to pick any color available (See Figure 5-4).

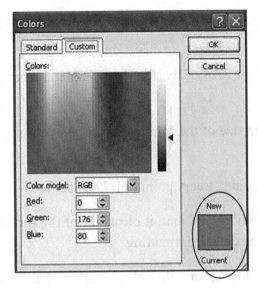

Figure 5-4

You can click anywhere inside the color choices to view the available colors. The bottom right side of the dialog box will allow you to see the current color (the bottom half of the square) and the new color you have clicked on (the top half of the square). If you like the new color all you have to do is click the OK button to make the change.

Click the Cancel button to leave everything as it is

Change the Expenses line to look like the Income line, only make the font color Red

Leave the last line as it is

There are a few other buttons in the Font Group that we have not mentioned (see Figure 5-5).

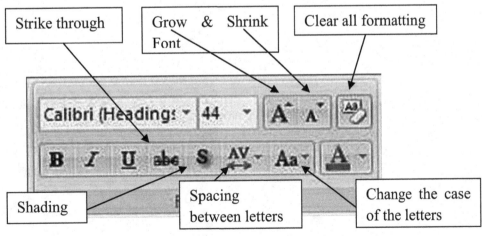

Figure 5-5

The Strike through command will put a line through all of the letters in the word(s) selected.

The shrink font will decrease the size of the font by one point every time you click the button.

The grow font will increase the size of the font by one point every time you click the button.

Note: Fonts are measured in points. Every point is 1/72 of an inch.

The clear all formatting button does just that, it clears all of the formatting from the selected text and returns it to the default formatting.

The shading button adds a shadow around the selected text to help it stand out from the rest of the text.

The spacing between letters button allows you to determine the spacing between the letters of the selected text. You can choose from five standard choices or make a custom spacing.

The change case button will allow you to change the case of the selected text to capital or lowercase.

Save your changes

Lesson 5 – 2   Using the Dialog Box

There are a lot of things that you can do from the Ribbon, but there are even more formatting things that we can do. These need to be done from the Font Dialog Box.

Open the Financial Meeting **presentation if necessary**

Select the text Profit / Loss on **slide 2**

Click the Font Dialog Box Launcher

The Font Dialog box will jump onto the screen and is shown in Figure 5-6.

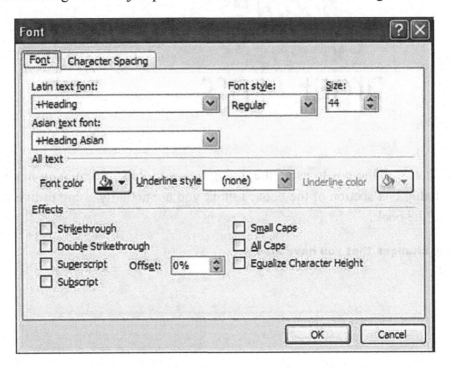

Figure 5-6

The top half of the dialog box has things that we have discussed using the Ribbon. We have not discussed all of the effects on the bottom part of the dialog box.

The Strikethrough effect we mentioned previously. It puts a line through the selected text. The Double Strikethrough effect puts two lines through the selected text. An example of this effect would be ~~Profit / Loss.~~

The Superscript effect moves the text slightly above the rest of the text and makes the selected text smaller than the rest of the text. An example of this would be $^{Profit\,/\,Loss}$

111

The Subscript effect moves the text slightly below the rest of the text and makes the selected text smaller than the rest of the text. An example of this would be ~Profit / Loss~

The Small Caps effect changes all of the letters in the selected text to capital letters with the first letter in each word being slightly larger than the other letters. An example of this would be PROFIT / LOSS

The All Caps effect will change all of the selected text into capital letters. An example of this would be PROFIT / LOSS

The Equalize Character Height effect will change the height of all of the letters to make them all the same height. An example of this is shown in Figure 5-7.

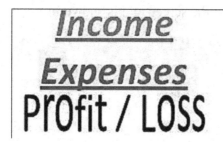

Figure 5-7

That is all there is to using the dialog boxes in PowerPoint. You made it through the dialog box section of the book. I think you are actually going to live through the rest of the book.

Save any changes that you have made

Lesson 5 – 3   Number and Bullet Lists

There may come a time when you want the viewer of your presentation to follow specific instructions. It would be great if the instructions were in a numbered fashion so they could be easily followed. In this lesson we will create a numbered list. We will also create a list using bullets instead of numbers.

Create a new blank presentation

This can be done by using the Office button and choosing new then blank presentation, or you can click the New shortcut on the Quick Access toolbar if you added it.

Type the following in the Title section on slide 1

The proper way to turn on your computer

Add a second slide

Click the mouse where it says "Click to add text"

Click the down arrow on the Numbered List button

This is located in the Paragraph Group which is on the Home Tab and is shown in Figure 5-8.

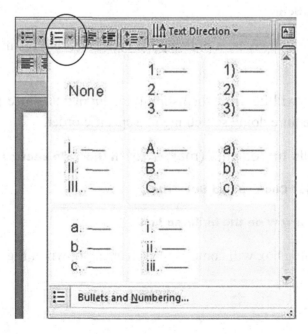

Figure 5-8

Click on the first choice, the one with a number and then a dot

> The number 1 appears on the slide. Now you can type the first thing that you want in your list.

Type Take a deep breath **and then press Enter**

> Immediately the number 2 will appear on the slide ready for you to type the second item in your list.

Type Exhale **and then press Enter**

> The number 3 will appear on the slide ready for you to type the third item in the list.

Type Press the On/Off button **and then press Enter**

> The number 4 will appear on the screen. We don't have a forth point to our list so we need to get rid of this number.

Press the Backspace button

> This will get rid of the number 4 and take the insertion point back to the left margin.

Save this presentation as Numbered List **and then close it**

> If the order of the items in your list is not important, a bulleted list might work even better that a numbered list.

Create another new Presentation

> In this presentation we will create a bulleted list using a recipe for making Almond Cinnamon Balls.
>
> Using the bulleted list will be great for listing the ingredients, since just listing the ingredients does not require doing something in a specific order.

Type Almond Cinnamon Balls Ingredients (makes 15) **in the Title section**

Click the mouse where it says "Click to add Sub Title

Click the downward pointing arrow on the bulleted List

> The Bullet Library dialog box will come to the screen as shown in Figure 5-9.

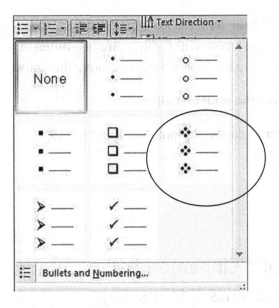

Figure 5-9

From here we can choose which bullet we want to use in our list. The bullet is the small dot or shape that we will have in front of every item in our list.

From the library choose the bullet with the four small diamonds by clicking on it

When you click on the desired bullet, the dialog box will disappear and the first bullet will appear in your presentation. The insertion point will be directly after the bullet.

Type 1 ½ Cups ground almonds **and the press Enter**

When you press the enter key the second bullet will appear, ready for you to type the next item in the list.

Add the following items to the list by typing each item and then pressing Enter

1/3 Cup granulated sugar

1 Tablespoon ground cinnamon

2 white eggs

Oil for greasing

Confectioner' sugar for dredging

If you are use to working with bulleted lists in Word or Excel, you know that pressing the Enter key twice in a row will remove the last bullet. This will not work in PowerPoint. You must use the backspace key to remove the unused bullet.

Press the Backspace key to remove the unused bullet

When you are finished, your slide should look like Figure 5-10

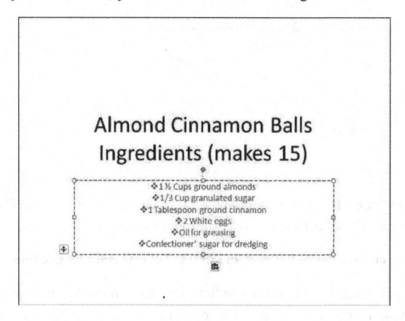

Figure 5-10

Well, it almost looks okay. In the next lesson we will see if we can make it look a little better. For this lesson it is just fine.

Save your work as Bullet List

Lesson 5 – 4   Paragraph Alignment

If you are one of those people who like everything nice and neat and even, this lesson just might make your day. In this lesson we will be discussing how your text is aligned on the page. We will also discuss line spacing and background shading.

If necessary open the Bullet List **presentation**

When we last looked at this presentation the text in the bulleted list was not very neat. Actually the text was all over the place. This did not look very professional. Let's see if we can fix that.

We will find the Paragraph Alignment commands in the Paragraph Group of the Home Tab (See Figure 5-11).

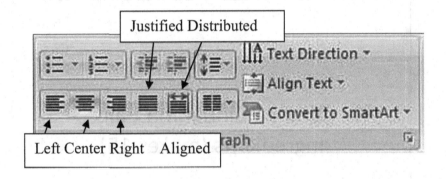

Figure 5-11

The text in the Title and Sub Title areas is by default center aligned. This means that the text starts in the center and works its way out toward the two sides at an equal pace. The result of this when there are several lines with different lengths will look like Figure 5-10.

In this lesson I will explain how the different alignment options will affect your slide's appearance.

The first thing we need to do is select the text that we are going to change.

Using your mouse select all of the text in the Sub Title area

There are five ways you can align the selected text. We have already discussed the Center Align option. The others are easy to figure out what they do if you click on the buttons. The left align will align the text on the left side of the slide. The right align button will cause the text to be aligned on the right side of the slide. The Justify button will keep both the left and right sides nice and even is the text covers more than one line (you need it to cover at least two complete lines to see the actual effect). The Distributed effect is a little different. This will spread the text out to cover the entire line, even if there is not enough text to cover the line. This all sound good, but it will be easier to understand if you actually use the commands.

Using the mouse click on each of the available choices starting with the left align button

After you have seen how each command affects your presentation click on the left align button

Your slide should now look like Figure 5-12, after you click outside of the text box.

Figure 5-12

Also in the Paragraph Group is button that has Align Text on it. This button will allow you to align the text in the middle of the row that holds the text. You can also choose to align the text just above center or just below the center of the row.

Directly above this is a command that has Text Direction on it. With this you can choose to change the direction of the selected text. The text can be horizontal (the default) or vertical or it can be rotated either by 90° or 270°.

These are the normal things you might want to change in the Paragraph Group. Feel free to select some text and play with these choices. You can always use the Undo button to remove any changes that you make.

Save and close the presentation when you are finished

Lesson 5 – 5   Using the Format Painter

The Format Painter allows you to copy the formatting from one section of text and apply the same formatting to another section of text. This could save you a considerable amount of time if you are constantly applying the same formatting for different parts of your presentation. In this lesson we will use the Format Painter to copy the formatting form one section of text and apply it to another section of text.

Open the presentation titled Financial Meeting

This is a presentation that we started earlier and now it is time to do some more work on it. We will copy the formatting that we applied to the words Income and Expenses and then we will apply that formatting to other text on other slides.

Click on slide 2 and select the line with Income on it

Now all we have to do is copy the formatting from here and paste it to another section of text.

With the text still selected click on the Format Painter

The Format Painter is located on the Home Tab in the Clipboard group (See Figure 5-13).

Figure 5-13

Before we can copy formatting, we have to let PowerPoint know what we are copying. That is why the text must be selected before we click on the Format Pinter.

As you move the mouse into the text area the pointer will be replaced by a small paint brush. To paint the formatting to other text click the mouse at the beginning of the text you want formatted and drag the mouse to the end of the text you want formatted.

Move the mouse pointer to the word Income on the third slide

You will need to click on the third slide to bring it to the top and then move the mouse pointer to the word Income.

Click the mouse and hold the left mouse button down

Drag the mouse to the end of the word

Release the left mouse button

Once you release the mouse button the painting will stop.

Just to make sure that we understand how this works, let's add another slide tour presentation.

Add a new slide to the presentation after slide three and make it a Title slide

At the top (in the Title section) type the word Expenses

In the Sub Title section type the following (be sure to press Enter after the first two)

> Maintenance
>
> Utilities
>
> Salaries

Using the technique you just learned with the Format Painter, copy the formatting from the word Expenses **on slide two to the word** Expenses **on slide four**

Note: If you have several items (or more than one) you want to paint with the Format Painter, double click the Format Painter instead of single clicking on it. Double-clicking the Format Painter will keep it turned on until you decide to turn it off. It can be turned off by clicking the Format painter again or by pressing the ESC key on the keyboard.

Save your changes

Chapter Five   Review

Formatting characters, fonts, and paragraphs can increase the impact your presentations have on the viewer.

Most of the common formatting can be done using the Font Group of the Ribbon. Additional formatting can be done using the Font Dialog Box.

If you need steps to be performed in a specific order, you will probably want to use a number list. If the order of the steps does not matter, a bullet list will be better.

Use the Paragraph Group to align the text inside a textbox on the slide.

The Format Painter will allow you to quickly copy the formatting from the selected text to other text.

Chapter Five   Quiz

1) Where, in PowerPoint, would you find the double strike through command?
2) What Tab and Group on the Ribbon has the Clear All Formatting command?
3) If you want to change the color of the selected text to one of the Theme or Standard colors, you just need to click on the desired color from the drop down list on the Font Color command. **True or False**
4) Fonts are measured in points. How large is a point?
5) Explain what effect the Equalize Character Height command has on selected text.
6) Explain why you would use a numbered list.
7) What is the difference between a left justified and a right justified paragraph?
8) The Format Painter is provided so you can copy the formatting from the selected text to other text. **True or False**
9) If the order of the items in a list is not important, which type of list should you use?
10) How do you get the Format Painter to stay activated until you are finished using it?

Chapter Six Using Themes and Styles

Imagine how great it would be if you could find an easy way to have all of your slides have that special look like you hired a professional firm to design them. Using themes will provide you with everything you will need to make your slides appear just like that. Themes (formerly called templates) will give all of your slides the same constant look throughout your presentation. There is more good news. If you save a theme that you have modified you can also use that theme in both Word 2007 and Excel 2007. This way the layout of all of your papers, spreadsheets, and PowerPoint presentations can all have the same look and feel.

Lesson 6 – 1   Applying a Theme

In this lesson we will apply a pre-made theme to one of our presentations.

Open the Financial Meeting **presentation**

This presentation is using the standard white office theme. This was a good starting point for us in PowerPoint and it did its job. Now we are much more experienced and need to move beyond just using the default theme.

Microsoft has put together a collection of color schemes that work together and provide your presentations with a professional look. You no longer have to spend hours trying to find just the right colors that look great together.

The themes are located on the Design Tab of the Ribbon (See Figure 6-1).

Figure 6-1

More Themes

Click the More Themes drop down arrow

The available themes will drop down and you can choose the theme that best suits your needs. This is shown in Figure 6-2.

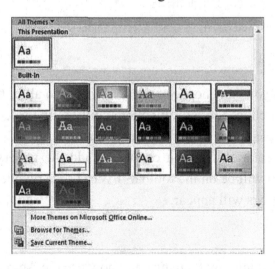

Figure 6-2

As with most of the new features you can see what the theme will look like in your presentation before you actually apply the theme. This is sort of new to Office 2007. In the older versions of Office there were a few different options that you could preview before you applied the changes, such as changing the font. If you selected some text and then clicked the drop down arrow by the font name, you could see a preview of the text with the new font. This feature was not available for the themes until Office 2007.

Move your mouse over the different themes and let the mouse hover over each one for a few seconds

As the mouse pauses at each theme the slide will change to reflect this theme. This way you can see the theme and decide if this theme will work for your presentation.

One of my favorite themes is the "Apex" theme.

If you want to apply this theme to all of the slides in your presentation, all you have to do is click on the theme. If, on the other hand, you do not want this theme applied to every slide you can right click on the theme and choose which slides have the theme applied to them.

Right-click on the Apex theme

A drop down menu appears and is shown in Figure 6-3.

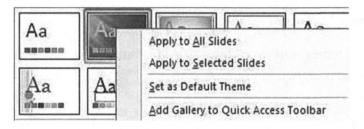

Figure 6-3

From the drop down menu you can choose to add this theme to all of the slides in the presentation. You can also choose to add this theme only to the selected slide(s). If you really like this theme you could choose to make this your default theme and use it for all of you new presentations. If you choose to add the gallery to the Quick Access Toolbar, a new icon will be on the toolbar. Every time you click on the icon a drop down menu of the themes will appear.

Click on Apply to all slides

All of the slides in the presentation will change and now use the Apex theme.

Save your changes

Lesson 6 – 2   Modifying a Theme

The theme feature is fantastic, but what if you only like part of the theme and not every detail of it? You might love the background of the Apex theme but you may not like the way the text looks. If you want, you can modify an existing theme and save it as your own theme.

Open the Financial Meeting presentation if it is not open

First let's look at the available color choices that are available for our chosen theme.

The color choices are on the Design Tab in the Themes Group.

Click on the Colors command

A menu will drop down showing all of the different color combinations that are available for our chosen theme. Again, you can use the live preview feature to see how each color combination will change the look of the presentation.

Move the mouse down the list of color combinations and see how they affect the presentation

You will notice that not all of the colors seem to be in the presentation. The first two colors in the group affect the background shading colors. The third color will affect the color of the text. The third, and the rest of the shown colors, will affect the color of the charts that are in the slides. You will see this when we insert a chart into our slide presentation.

Click on the Module color combination

I think that I like this a little better than the default color that comes with Apex.

What do you think about changing the font to a different style?

Click the Fonts command in the Themes Group

As you can see, the default font is Lucida Sans (it is the one highlighted). Again the live preview is available. You can scroll down the list of theme fonts and see how they will look in your presentation.

Move the mouse over each font choice and watch your font on the slide to see which one you like

I like the Trek font, so let's use it.

Click the mouse on the Trek **font**

Normally I would suggest that we change the effects on the slide, but they are extremely difficult to see any change at this point. We will go back and look at the effects when we insert a chart into our slides. The differences will be a lot easier to see then.

Let's pretend that we really like the changes that we have made. We might as well pretend that we like them so much that we will want to use them again. Hey! Why don't we save these changes so we can use them again?

Click the More Themes **arrow**

The drop down menu will again drop down onto your screen. Figure 6-4 shows you the menu.

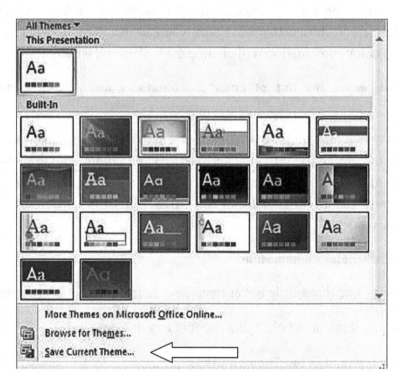

Figure 6-4

Click on Save Current Theme

When you click this option a dialog box will come to the screen asking for you to name the theme. When you type in a name and click the Save button, the theme will be saved and you can use it whenever you need it. The theme will be available for your Word documents, Excel spreadsheets, and PowerPoint presentations. Figure 6-5 shows the Save Current Theme Dialog box.

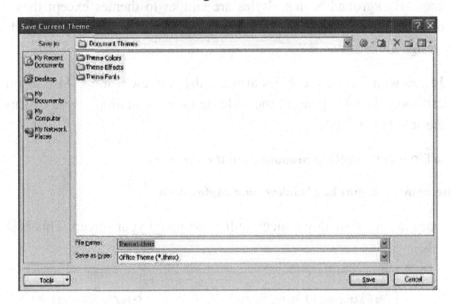

Figure 6-5

In the file name textbox type My_Apex_1.thmx

There will be a new icon for our theme. You should be able to find it in the Themes Group. That is all there is to modifying a theme and saving it with a new name.

Save your changes

Lesson 6 – 3   Adding Background Styles

In the previous two lessons we learned about themes in this lesson we will learn about Background Styles. Styles are similar to themes except they only affect the background of the slide. The styles are found in the Background Group on the Design Tab.

Just as with themes the styles also use the preview feature. Also just like themes you can apply the style to all of the slides in the presentation or you can apply the style to the selected slide(s).

Open the Financial Meeting **presentation if necessary**

Click the command that has Background Styles on it

Once again a drop down menu will show up on your screen. This is shown in Figure 6-6.

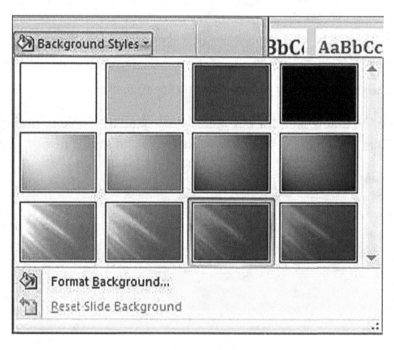

Figure 6-6

You can tell the current style by the gold box around it. With the preview feature you can move your mouse over the different background styles to see the effect they have on the slide.

Move your mouse over and let it pause on each background style

If you find a background style that you like, you can click on it to add it to all of the slides in the presentation. If you only want to change the background on one slide you can right-click on the style and choose apply to selected slides.

I am not sure that I like any of these choices, so let's try something different.

Click on Format Background

This will bring the Format Background Dialog Box to the screen. From here we can make our own choices about how the background should look. Besides we know more than everyone else, right?

This also uses the live preview for some of the effects.

First click the radio button on Solid Fill, **then on the** Gradient Fill, **and then on** Picture or Texture Fill

That is the one I like. Let's use it.

Click the down arrow of the Texture **command**

This will bring all of the textures we can use to fill the background of the slide to the screen. The live preview is not available for this part of the dialog box. To see what each one will look like you will have to click on it.

After you have tried all of the textures, click on and choose the one in Figure 6-7

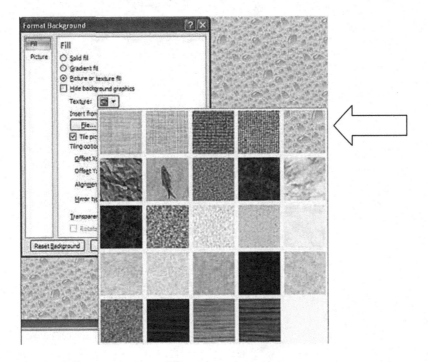

Figure 6-7

If you would prefer to have a picture as the background instead of the texture, you would need to click on the Insert from choices of: from file, or the clipboard, or from clipart buttons. You cannot have pictures and texture both as a background, so you need to decide which one you want.

Click the Close button

At this point we will not apply this to all slides, as a matter of fact we are not even going to keep this style.

Click on the Background Styles command again and this time click on Reset slide Background

Everything is back to the way it was before we started playing.

Save your changes

Chapter Six   Review

Using themes and styles will give all of your slides the same layout and consistency.

PowerPoint has several pre-made themes available for you to apply to your slides. Microsoft has spent a lot of time putting together a collection of colors that work together.

Themes use the live preview feature so you can see how the theme will look before you apply it.

Themes can be applied to all of the slides in the presentation or only to the selected slides.

All themes can be modified to fit your needs.

Background styles are similar to themes, but they only affect the background of the slides. Background styles can also be applied to all slides or only the selected slides.

Chapter Six   Quiz

1) Saved themes can be used in Word 2007, Excel 2007, and PowerPoint 2007. **True or False**
2) The Themes Group is located on what Tab of the Ribbon?
3) The only way to see how a theme will look in the presentation is to apply the theme to at least one slide in the presentation. **True or False**
4) If you click the mouse on one of the pre-made themes, the theme will be applied to which slide(s)?
5) Pre-made themes are just that, pre-made, and cannot be modified. **True or False**
6) Background Styles use the Live Preview feature. **True or False**
7) You can select both Gradient Fill and Texture Fill when formatting the background of a slide. **True or False**

Chapter Seven — Working with Tables

Putting a table into your presentation will allow you to arrange information in a neat and organized grid. A table contains cells, where the text (data) is displayed, and the cells are arranged in columns and rows. You can format a table, such as giving it borders as well as shading and coloring options. As you will see, tables are very powerful but few people know how to use them, so they don't. In this chapter you will learn how to use tables.

Once you learn how to use tables, you might wonder how you have survived this long without using them. With a table you can:

Align text and numbers: Tables make it easy to align text and numbers in rows and columns.

Create a form: You can use tables to store lists names, addresses and phone numbers.

You can copy and paste a table's information into a Microsoft Excel spreadsheet.

Create a publication: Tables allow you to create calendars, brochures, business cards, and many other publications.

Lesson 7 – 1 Creating a Table

In this lesson you will learn how to create a table and insert information into it. The first thing you need to do is insert a table into your presentation.

Create a new blank presentation

Change the slide to Title and Content

Changing the slide type will allow us to insert a table into this slide. Your slide should look like Figure 7-1.

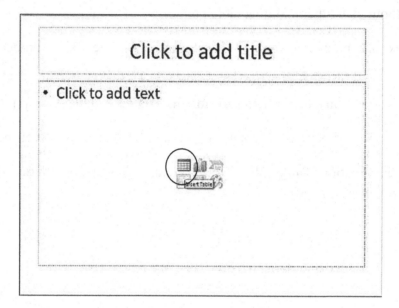

Figure 7-1

Click on the Insert Table **icon**

When you click on the Insert Table icon the Insert Table Dialog Box will come to the screen. Figure 7-2 shows this dialog box.

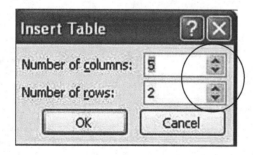

Figure 7-2

For our table we are going to have 2 columns and 9 rows.

Using the up and down arrows change the setting to reflect 2 columns and 9 rows

Click the OK **button**

Your slide should look like Figure 7-3.

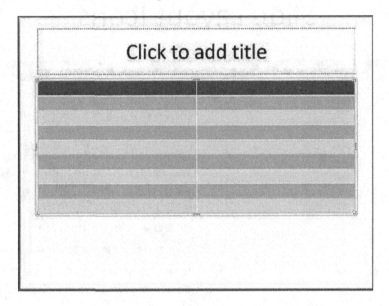

Figure 7-3

As you can see, every other row is highlighted. This way the table contents are easily separated. Now all we have to do is add the information to our table.

In the Title section add the following text

> Slide Layout Icons

Click in the top row on the left side and add the following text

> Icon Type

Press the TAB **key and add the following text to the right side**

> Description

Press the TAB **key again to move to the left side of the second row and add the following text**

> Title

Press the TAB **key again to move to the right side of the second row and add the following text**

> Inserts a title or heading

Fill in the rest of the table so that it looks like Figure 7-4

Slide Layout Icons

Icon Type	Description
Title	Inserts a title or heading
Text	Inserts text or lists (bulleted or numbered) into the slide
Contents	Allows you to select the type of content you want to insert into the slide. You can choose from Table, Chart, Smart Art Graphics, Picture, Clip Art, or Media Clip
Table	Inserts a table into the slide
Chart	Inserts a chart into the slide
Smart Art Graphics	Inserts smart art into the slide. This can be one of several types including flow charts and block lists
Picture	Inserts a picture from a file into the slide
Clip Art	Inserts clip art into your slide

Figure 7-4

Oops we have one more thing to add to the table. We don't have a place for Media Clips. No tears, we can do this. If we have our insertion point inside the last cell of the table we can use the TAB key to add another row.

With the insertion point inside the last cell of the table press the TAB **key**

Like magic another row appears. Now we can add the last item to our table.

Add the last item to the table as shown below

Media Clip	Inserts a music, sound, or video clip into the slide

Figure 7-5

Well we finally have a table in our slide. Somehow it just doesn't look quite right. Perhaps if we adjusted the table so that the left side is smaller than the right side it would look more normal.

Move your mouse to the center line between the cells in the top row

When your mouse gets to the correct position the pointer will turn into a small double vertical line with an arrow pointing to the right and left. When it get to this point you need to click and hold the left mouse button down and then drag the mouse to the left.

Click and hold the left mouse button down and then drag the mouse to the left

The center line will start moving along with your mouse and the cells will change size when you release the left mouse button.

When your table looks similar to Figure 7-6 release the left mouse button

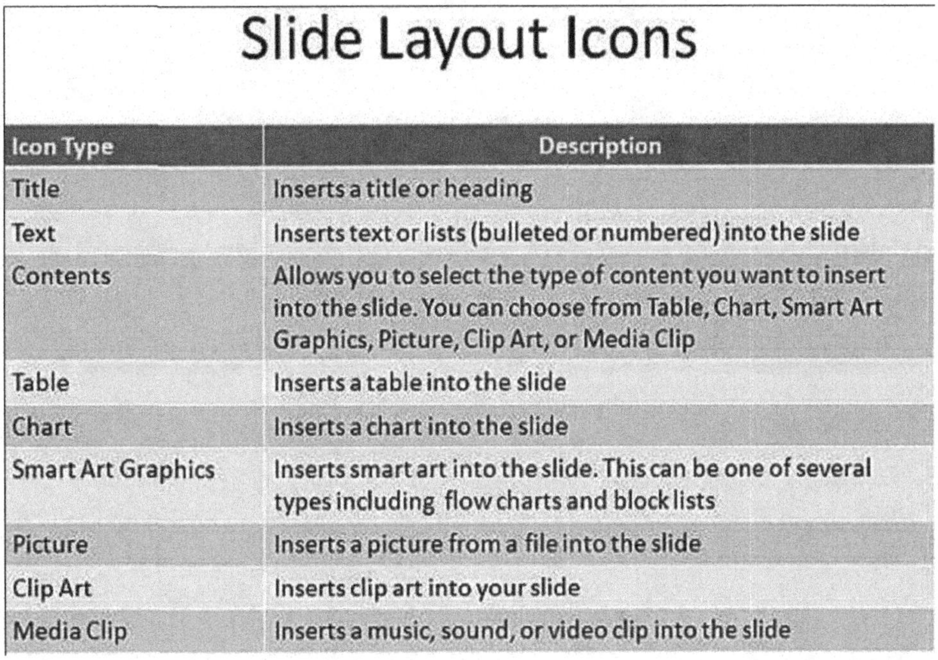

Slide Layout Icons

Icon Type	Description
Title	Inserts a title or heading
Text	Inserts text or lists (bulleted or numbered) into the slide
Contents	Allows you to select the type of content you want to insert into the slide. You can choose from Table, Chart, Smart Art Graphics, Picture, Clip Art, or Media Clip
Table	Inserts a table into the slide
Chart	Inserts a chart into the slide
Smart Art Graphics	Inserts smart art into the slide. This can be one of several types including flow charts and block lists
Picture	Inserts a picture from a file into the slide
Clip Art	Inserts clip art into your slide
Media Clip	Inserts a music, sound, or video clip into the slide

Figure 7-6

There is only one last thing we need to do to our table. We need to fix where the word Description is located.

Click the mouse inside the cell where the word Description is located and then click the Center Align **button**

I am sure that you remember that the Center Align button is located on the Home Tab in the Paragraph Group, and I probably didn't even need to mention it.

Just when I thought you were finished with tables I noticed that there is a new Tab on the Ribbon.

I guess that we better have another lesson or two.

Save the presentation as Table Example

Lesson 7 – 2   The Table Tools Design Tab

Open the Table Example **if it is not open and click the mouse inside the table**

The Design Tab of the Table Tools was not visible before we added a table to our slide. If we click our mouse outside of the table this section goes away. It will only come back if we click the mouse inside of the table. Now would be a good time to look at the various groups that are available on this tab.

Click on the Design Tab

The first group is the Table Style Options Group and is shown in Figure 7-7.

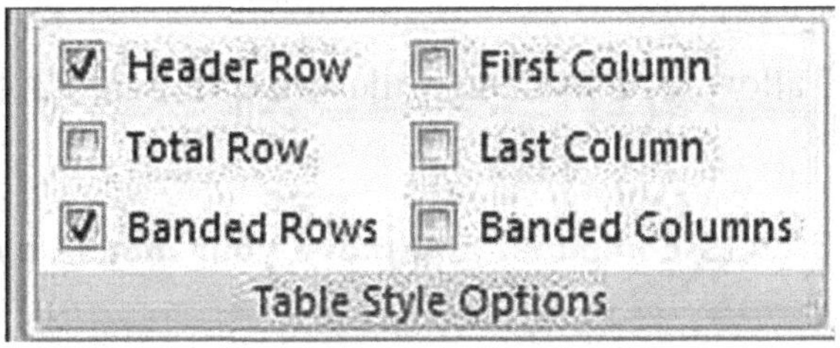

Figure 7-7

The first option is the Header Row option. This option is checked by default.

Click the checkbox next to Header Row to turn it off

Click the checkbox next to Banded Rows to turn it off

As you can see these make a big difference in the look of your table.

Click the checkbox next to First Column and then click it again to turn it back off

Click the checkbox next to Last Column and then click it again to turn it back off

Click the checkbox next to Banded Columns and then click it again to turn it back off

Click the checkbox next to Header Row and Banded Rows to turn them back on

The next group is the Table Styles Group and is shown in Figure 7-8.

Figure 7-8

This group will change the style of the table. This is a lot like the themes we talked about earlier. The styles also use the live preview feature. This means that you can run your mouse over the different styles and see the results on your slide.

Move your mouse over the different styles and watch the results

More styles, a lot of them, are available if you click the More Styles down arrow.

Click the down arrow on the Shading **command and move your mouse over the choices to see how the different colors will look in the table**

The Borders command will bring a drop down menu showing all of the available borders you can add to the table. There is no live preview for you to see the results without adding a border. You also have to click on the down arrow to see the different borders.

The Effects button will allow you to change how the individual cells look in the table.

Click the down arrow on the Effects **command and move your mouse over the different choices**

The effects menu does have the live preview so you won't have to actually add the effect to see how it looks.

The next group is the Word Art Styles Group and is shown in Figure 7-9.

Figure 7-9

This group will allow you to add effects to the selected text inside a cell, or to all text in the table.

Select the text in the top left cell and then click on the Quick Styles **icon**

Run your mouse over the different WordArt styles and see how each one changes the way the text looks

Repeat this process for the Text Fill **command and then the** Text Outline **command and lastly the** Text Effects **command**

Each one of these will change the way the text looks in your slide.

The last group on the Design Tab is the Draw Borders Group. This is shown in Figure 7-10.

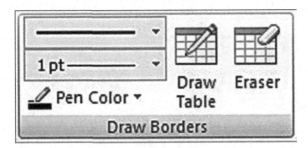

Figure 7-10

This is not an area where I excel and I am not even going to think about showing you the borders that I drew. But I will tell you how it works. You can play and try your hand at drawing borders.

The good news is that even if you cannot draw a straight line, your borders will look great. It is really not that hard.

The box on the top left, the one with the straight line on it, is the pen style. This will let you decide what type of border you want to draw. It could be a solid line, or a dashed line, or even a line with asterisks.

The next box down will decide how thick the border will be when you draw it.

The next box is the pen color, and you can choose what color the border will be.

If you click on the Draw Table button and move your mouse inside the table the pointer will turn into a pencil. To draw all you have to do is press and hold the left mouse button and drag it along the borders of the cell(s). As you drag it PowerPoint will make a dotted line on the border and all you need to do is make sure your mouse starts at one end of the dotted line and ends at the other end of it. Even if your mouse wiggles all the way down the edge of the cell, it won't matter. PowerPoint looks at the beginning and ending points.

Change the color to Red and click on the Draw Table **command and draw a border around the three inside borders of the cell that has the Contents in it. Do each border one at a time**

It is, at least for me, impossible to draw a decent border on the outside edge of the table. You can try if you want, but I don't think you will be able to come up with one solid line for a border. If you can, my hat is off to you.

Save your changes

Lesson 7 – 3   The Table Tools Layout Tab

Open the Table Example **if necessary and click the mouse inside the table**

The Table Tools Layout Tab has seven groups on it. We will look at these groups in this lesson.

Click on the Layout Tab

The first group on the left is the Table group and is shown in Figure 7-11.

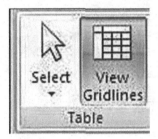

Figure 7-11

This group has two commands in it; the Select command and the View Gridlines command.

If your cursor is inside of a cell you can click on the select command and then decide if you want to select the entire table or the column that the cursor is in or the row the cursor is in. This command will not select just the cell or the text in the cell. You must choose the entire table or the column or the row.

Click the mouse in the cell that has the word Title **in it**

Click the select button and try the different choices

The next command is the View Gridlines command. Clicking this button will toggle between seeing the gridlines and not seeing them. You might very well spend several minutes clicking on this button and wondering why I lied to you. Well I didn't lie, but it is very hard to see this with the existing table. To make this easier on you we are going to add another table to our presentation and you will be able to see what this command does.

Add a new slide and insert a table into the slide

In the table have 11 columns and 11 rows

145

We want this to be even easier to see, so we are going to change the table style. We discussed styles earlier and now we are going to change this table to a different style which will show the gridlines a little better.

Click inside the table and then click in the Design Tab that is in the Table Tools Tab

Click on the "More" arrow that is in the lower right corner where the styles are shown

This will cause the large drop down menu to come onto the screen. We will be using the dark blue that is shown in Figure 7-12 with the gold border around it.

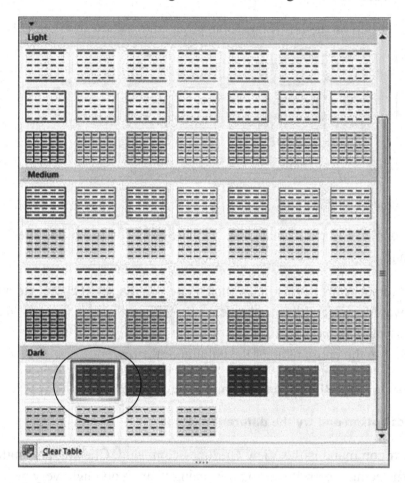

Figure 7-12

This particular style will show the gridlines better than the default style.

Click on the dark blue style

Go to the Layout Tab and click on View Gridlines at least twice

This is what the View gridlines command does.

146

The next group is the Rows and Columns Group and is shown in Figure 7-13.

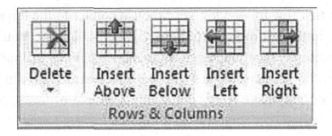

Figure 7-13

The first command is pretty obvious. If the insertion point (the flashing vertical line) is inside a cell and you click the Delete button you can choose to delete either the row or the column.

The next two commands deal with inserting rows. If the insertion point is inside of a cell you can click the Insert Above command to insert a row above the row that has the insertion point in it. If the insertion point is inside of a cell you can click the Insert Below command to insert a row below the row that has the insertion point in it.

The next two commands work the exact same way only with columns. You can insert a new column to the left or the right of the column that has the insertion point in it.

The next group is the Merge Group and is shown in Figure 7-14.

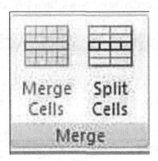

Figure 7-14

The first command in this group will let you merge two or more cells together to make one cell. Why would you ever want to do something like that? I can think of one reason. You might want to have one cell with all of the heading in it. This way you would not have a separate cell for each heading. Let's try it and see what happens.

Click on the first slide to bring it to the screen

Select the first row of the slide

You can select the first row by moving your mouse pointer to the top row and while keeping the mouse in the left margin click the left mouse button. You can tell when the mouse is in the correct position because the mouse pointer will turn into a small black arrow. When the pointer changes into the black arrow all you have to do is click the mouse. You may have to try doing this a couple of times, as it may be a little tricky.

Click the Merge Cells **command**

The result might be a little surprising, and I can hear you now saying "Bill what have you done to us"? Why would I want something like this in my presentation? Let's see if we can make it look a little better. The top of your presentation should look like Figure 7-15.

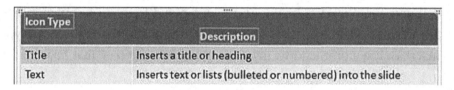

Figure 7-15

Click your mouse after the word Type on the first line

Press the Delete key on the keyboard

Press the Space bar on the keyboard until your presentation looks like Figure 7-16

Icon Type	Description
Title	Inserts a title or heading
Text	Inserts text or lists (bulleted or numbered) into the slide
Contents	Allows you to select the type of content you want to insert into the slide. You can choose from Table, Chart, Smart Art Graphics, Picture, Clip Art, or Media Clip
Table	Inserts a table into the slide
Chart	Inserts a chart into the slide
Smart Art Graphics	Inserts smart art into the slide. This can be one of several types including flow charts and block lists
Picture	Inserts a picture from a file into the slide
Clip Art	Inserts clip art into your slide
Media Clip	Inserts a music, sound, or video clip into the slide

Figure 7-16

The next command is the Split Cells command. This command will allow you to split one cell into two cells. There goes another one of those why questions. To find out let's continue.

Add a third slide to your presentation

Insert a table with 4 rows and 6 columns

We will use this to include some names and addresses of our colleagues. We will add the first and last name, address, phone number, city, state, and zip code. At the top we will have a row with the row heading on it. In general make your slide look like Figure 7-17. You will need to adjust the width of some of the columns for the text to fit on one row.

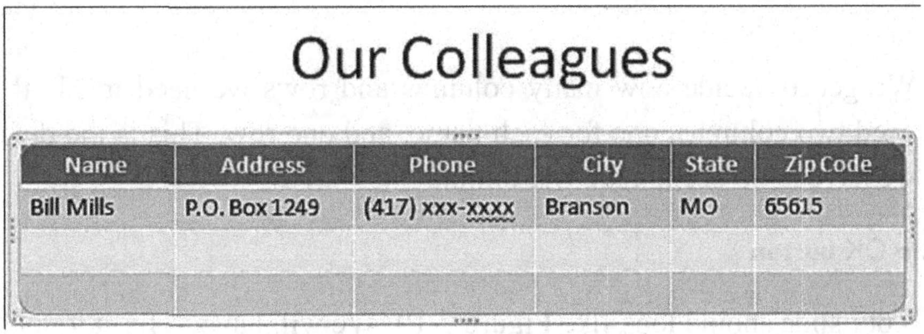

Our Colleagues

Name	Address	Phone	City	State	Zip Code
Bill Mills	P.O. Box 1249	(417) xxx-xxxx	Branson	MO	65615

Figure 7-17

You didn't really think that I would put my phone number in there did you?

If you want you can add a couple of other names in the table

Now that we are finished, we have decided that it would be nice to separate the first name from the last name in column 1. To do that we will need to use the Split Cells command. This command will not actually split the words in the cell for you, you have to do that. The command will split the cell into columns and rows. You have to move the data around to fit your needs.

Click on the cell with my name in it

Click the Split Cells **Command**

A dialog box will come to the screen as shown in Figure 7-18.

Figure 7-18

We get to decide how many columns and rows we need inside this cell. We only need two columns, one for each name, and one row. This is the default setting so all we have to do is click the OK button.

Click the OK button

Your slide should look like Figure 7-19. We will have to fix it to look like we want.

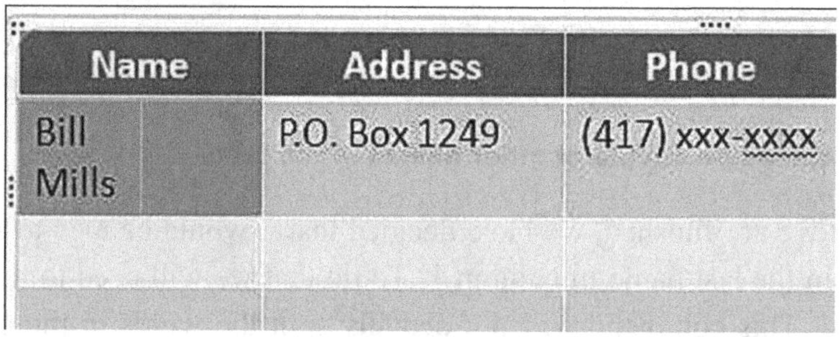

Figure 7-19

Making this look the way we want is a two step process. First we have to delete the last name in the first column and then add the last name in the second column.

Click the mouse after the word Bill and then use the Delete button to delete the last name

Click in the second column of the first cell and add the last name Mills

Your slide should now look like Figure 7-20.

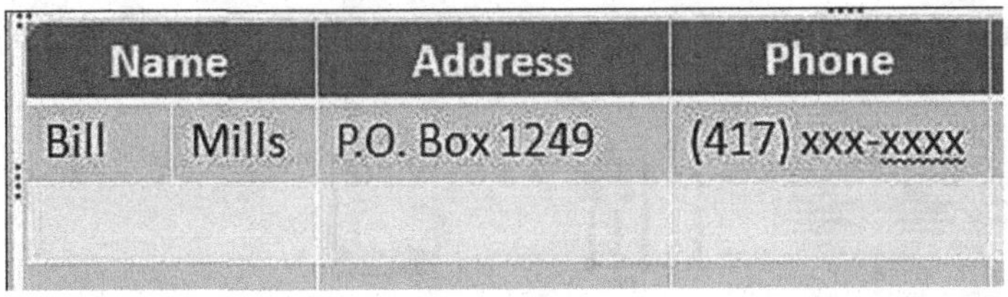

Figure 7-20

Note: You could have used the cut and paste commands to move the data instead of using the delete key and re-typing the text.

The next group we will look at is the Cell Size group. This group will allow you to change the size of each individual cell in the table. It will also allow you to make sure all columns and rows are the same size. This group is shown in Figure 7-21.

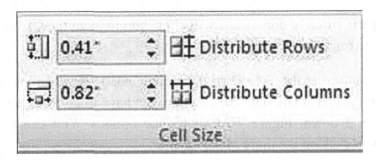

Figure 7-21

You can play with this if you want to, but I am only going to tell you how it works. If you select a cell (the insertion point will be in it) you can use the up and down arrows to adjust the height and width of the cell. You can also select an entire column or row.

Note: If you select the column, you can adjust both the height and width of each cell in the column. If you select a row you can only adjust the height of the cells, not the width.

The Distribute Rows command will make sure that all of the rows in the table are the same height.

The Distribute Columns command will make sure all of the columns are the same width.

Note: If you use either of the Distribute commands you will have to use the Undo function to get the rows and columns back to the original size. Clicking the command a second time will not bring the previous heights and widths back.

The next group is the Alignment Group and is shown in Figure 7-22.

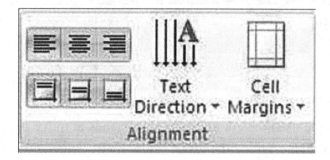

Figure 7-22

The top left group of commands will allow you to align the text either to the right, to the left, or in the middle of the cell.

The bottom left group of commands will allow you to align the text at the top, or the bottom, or the center of the cell.

The next command is the Text Direction command. This will allow you to change the direction the text is laid out. The text does not have to go from left to right; it can also go up and down. Figure 7-23 shows the options. You can also change the margins if you click on the more options button.

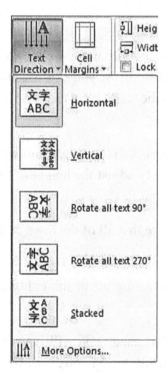

Figure 7-23

The next group is the Table Size Group and is shown in Figure 7-24.

Figure 7-24

These commands will let you adjust the height and width of the table as a whole, not just the cells. If you click the Lock Aspect Ratio checkbox, the height and width will both change when you change either the height or width.

The last group is the Arrange group and is shown in Figure 7-25.

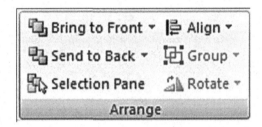

Figure 7-25

We will cover this later when we discuss inserting pictures into your presentations, but since it is relevant here I will give you a brief description of these commands. In Figure 7-26 and 7-27 I have inserted a picture into the slide and am using it as a background for the slide. I can bring the table to the front or send it back behind the picture using the Bring to front and the Send to back commands.

Figure 7-26   Figure 7-27

153

The Align command allows you to align the edge of several objects so they are all nice and even. The Group command allows you to group several objects together so that can be treated as one object. The Rotate command allows you to flip or rotate objects. You will understand these better when we get to the lessons on graphics.

Now we are finally finished with tables.

Save your presentation and then close it

Chapter Seven   Review

Tables will allow you to arrange information in a neat and organized grid of columns and rows.

Tables are normally inserted into specific slides, but can be inserted into any slide using the Table command that is on the Insert Tab.

You can set the number of columns and rows when you insert the table into a cell. Rows can, however, be added by pressing the Tab key when you are in the last cell. You can also right-click on the table and insert columns and rows. You can delete columns and rows by right-clicking on the table and choosing the delete options.

You can change the borders of a cell by dragging them to the new location.

Two new tabs are available when you are working with a table; The Design Tab and the Layout Tab. The Design Tab works with the available styles and borders, while the Layout Tab mainly deals with rows, columns, and cells.

Chapter Seven   Quiz

1) You can only pit a table in a Title and Content slide. **True or False**
2) Explain one way of adding a new row to a table.
3) How do you move a border of a cell?
4) What affect does unselecting the Banded Rows option that is in the Table Styles Options Group have on a table?
5) If you click the mouse inside a cell and then click the Select command that is in the Table Group of the Layout Tab, what choices do you have?

Extra Credit:

Name two different Tabs and Groups where you can center align text in a cell.

Chapter Eight   Working with Graphics

At the end of the last chapter I mentioned inserting a picture into a slide. A picture is only one of the many graphics that can be inserted into a slide. In this chapter we will insert pictures, text boxes, word art, and charts into our slides.

Graphics will make your presentations more exciting and the viewer will retain its contents longer if it your presentation is more pleasurable to watch. I don't have the statistics to back up what I just said so I will correct my statement to say that it is my opinion that the viewer will retain its contents longer if it your presentation is more pleasurable to watch. I think you will find out that my statement is accurate.

Let's have some fun with our slides.

Lesson 8 – 1   Inserting Pictures

Since we are starting something new, let's create a whole new presentation.

Create a new blank presentation

This can be done by clicking the Office button and selecting New and then Blank Presentation and then clicking the Create button. It can also be done by clicking the "New" button on the Quick Access Toolbar if you added it to the toolbar.

Type the following in the Title:

Branson Homeowners Association   [Press Enter Here]

Goes Across the U.S.A.

Change the font to Times New Roman

Change the font size to 36

Add the effect for Small Caps

Center Align **the text**

These changes can be made by launching the font dialog box (except the center-align which must be done from the paragraph group). To launch the dialog box, click the small button on the right side on the font group of the home tab. Your first slide should look like Figure 8-1.

Note: If you want to get rid of the box that has Click to add sub title in it. Click inside of it and then press the spacebar.

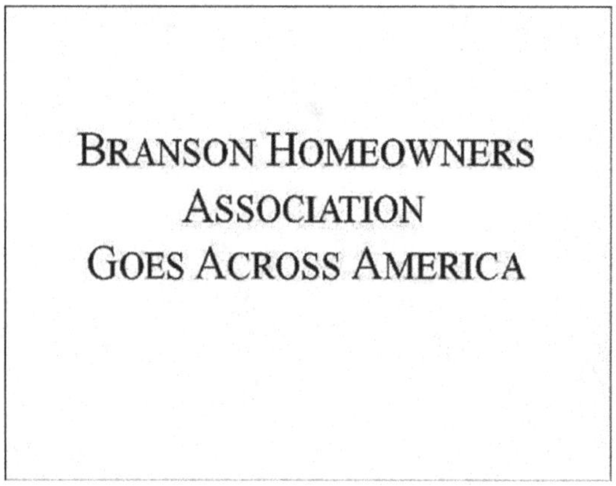

BRANSON HOMEOWNERS
ASSOCIATION
GOES ACROSS AMERICA

Figure 8-1

The last thing we want to do for this page is to give it a background picture. We want the background to be almost see-through. For this effect we need to use a watermark. PowerPoint 2007 will allow us to insert a picture or clip art and change the relative lightness (brightness) so that it is almost see through.

On the Design Tab click the dialog box launcher in the Background Group

This will bring the Format Background dialog box to the screen as shown in Figure 8-2.

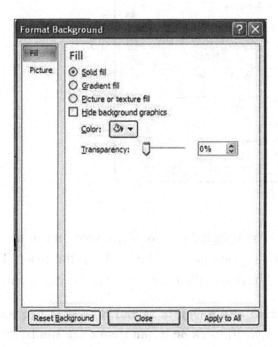

Figure 8-2

The first thing that we have to decide is what kind of background we want in our slides. For this presentation we want a nice happy little picture for our background.

Click on Picture or texture fill

Some additional items will be added to the dialog box and are shown in Figure 8-3.

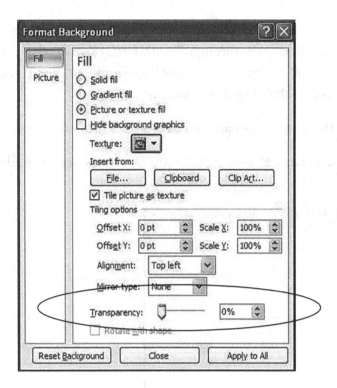

Figure 8-3

We are going to use a ClipArt image, but it will not show up in the choices if you click on the ClipArt button. If you insert a ClipArt from the Insert Tab, this image will be one of the choices. I have included this image in the downloaded files.

Click on the Insert from file **button and navigate to the files that you downloaded and then choose the image named** Sun.wmf. **Click on it and then click the** Insert **button**

The bright picture of the sun will be inserted as the background for the slide. The only problem is that the picture is so bright it is hard to see the text on the slide. We need to change the transparency of the picture. We want the picture to be almost see-through.

Increase the transparency until it reaches 80%

This should make the background almost see-through. Now let's make the picture a little bit smaller so it will fit on the slide.

Change the Offsets **so that they match the setting in Figure 8-4**

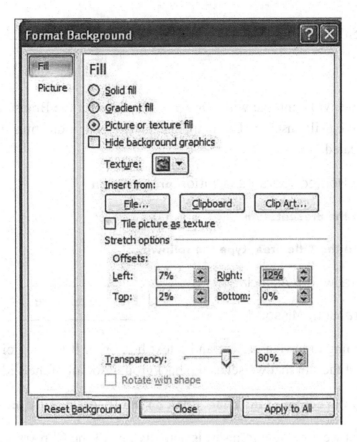

Figure 8-4

Now all we have to do is apply the background to every slide in the presentation.

Click the Apply to All **button**

Click the Close **button**

Every new slide that we create for our presentation will now have this in the background. You will also notice that there is a new theme on the Design Tab. We can apply this theme to other slides by simply clicking on it after we have a slide selected.

Save this presentation as Homeowners Association

Lesson 8 – 2   Working with Text Boxes

In this lesson we will continue with our series of slides for the Branson Homeowners Association. We will insert a US map and a text box indicating where Branson Missouri is located.

If necessary open the Homeowners Association **presentation**

Add a second slide to the presentation

On the second slide, in the Title area, type the following:

Branson Homeowners Association

Located in Branson, Missouri

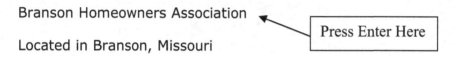

This slide is going to have a picture and a text box on it. It is also going to have the faded picture of the sun that we set as the background for all of the slides.

There are six icons in the center of the slide. Click on the Insert picture **icon**

The picture that we are going to insert is actually one of the ClipArt pictures that you have to go online to find. I am enclosing it in the downloaded files for your convenience.

Navigate to and select the US Map.wmf **picture**

Click on Insert

Your slide should look like Figure 8-5.

Figure 8-5

The picture of the US Map will be inserted into our document. Now all we have to do is mark where Branson is located and this page should be done.

On the Insert tab of the Ribbon, click on Text Box in the Text group

The mouse pointer will change into an upside down plus sign (+).

Move the mouse until the plus sign is somewhere around New York and click and hold the left mouse button down and drag the mouse to the right and down until the box is about the same size as the one in Figure 8-6.

Figure 8-6

Type Branson Missouri **inside the text box**

The text box will expand automatically to allow the words to fit inside it. It will expand vertically not horizontally. When you click outside of the box the border will disappear and the words Branson Missouri will be left on the slide. This might be okay, but this is not where Branson Missouri is located. We need something to let the viewer know where Branson is on the map. It would be great if we had an arrow pointing to where Branson is on the map.

On the Home Tab and in the Drawing Group is a line with an arrow on one end of it. We can use this to point to the location of Branson.

Click on the pointing arrow in the Drawing group

The mouse pointer will turn into a plus sign and the arrow will turn an amber color to indicate that it has been selected.

Move the pointer to the edge of the Textbox and click and hold the left mouse button down while you drag the pointer to the southwest part of Missouri and then let go of the left mouse button

Your slide should look like Figure 8-7.

Figure 8-7

The circles at the ends of the pointing arrow are there for you to use to move or resize the line. If you need to move the line you just click on one of the circles and hold the mouse button down and re-drag the line to a new location. When you click outside the area the circles will go away.

I personally don't care for the blue color of the line. If you right-click on the line you can make some changes to the default settings for the line.

Right-click on the line and choose Format Shape **at the bottom of the menu**

Click on the Line Color **button on the left side and change the color to** Red **from the drop down menu (See Figure 8-8)**

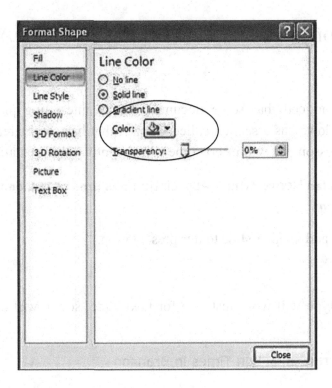

Figure 8-8

Now the line will stand out on the map. Your viewer will now be able to see where Branson is located. What is even better is now you know how to insert a textbox into a slide.

Save your changes

Lesson 8 – 3 Working with ClipArt

You might have noticed that some documents have interesting pictures, almost works of art. Windows has a section called ClipArt that has some great pictures for us to use. In this lesson, we will insert some ClipArt into our presentation.

Open the presentation called Homeowners Association **and press CTRL and End to get to the end of the presentation**

From here we will add another slide to our presentation.

Add a new slide

The default slide type will work just fine for next slide, so we will use the default slide layout.

Type the following in the Title area: Fun Times in Branson

We need to add some ClipArt showing how much fun you can have in Branson.

Click on the ClipArt button

This will bring the ClipArt pane to the right side of your computer screen. This is where you can search through all of the ClipArt to find the one that will suit your purposes. By the way you can search online through a couple of hundred thousand ClipArt images if you can't find one that you like in the built-in clips. We will search for an image that will show things we can do in Branson.

Click your mouse in the search box at the top of the ClipArt pane

Figure 8-9 shows the clip art pane.

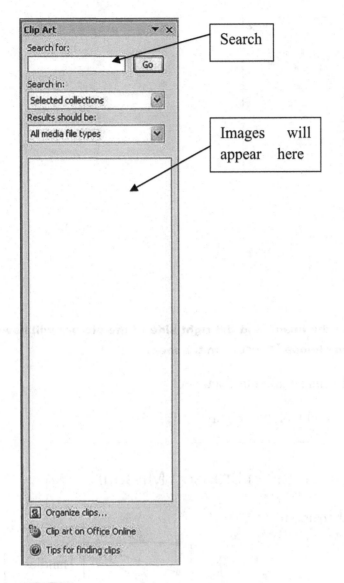

Figure 8-9

Type golf and then click on the Go button

In a few moments a, or several, small images should appear in the blank space. We should be able to find a golf picture we like.

Scroll down, if necessary, until you find the image of a female golfer shown in Figure 8-10

Figure 8-10

Move your mouse over to the image and the right side of the picture will have an arrow on it. Click on the arrow and choose Insert **from the menu**

This will insert the image into our document.

Your document should look like Figure 8-11.

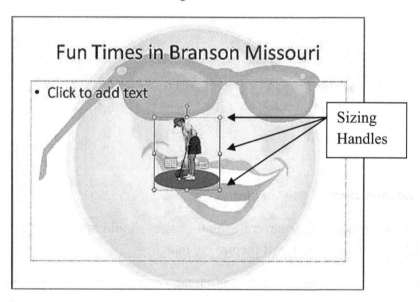

Figure 8-11

With the picture still selected, move your mouse to the lower right corner until your pointer changes into a double sided arrow at an angle.

The small white squares and circles are called sizing handles and they let you change the size of the Image by clicking and dragging the image.

Click and hold the left mouse button and drag the sizing handle down and to the right until the picture almost fills the entire page then release the mouse button.

The last thing we need to do is get rid of that annoying "Click to add Text" that is in the slide.

Click inside the "Click to add text" box and then press the spacebar to get rid of the text

Just so that you know, the "Click to add text" will not show up in the slide show when you run it. It is only visible when you are working on the slide, but it is annoying to have it there.

Your slide should now look like Figure 8-12.

Figure 8-12

Save your work when you are finished

Lesson 8 – 4 Positioning Pictures

Have you ever wondered how some people have positioned their picture at strange angles? Not all pictures in your presentations have to be aligned straight up and down. Pictures can be tilted to give your presentations a real professional look. In this lesson we will practice tilting the pictures to different angles.

Open the Homeowners Association **presentation and press CTRL and End to get to the end of the presentation**

Add a new slide and use the default layout

Add the following text to the Title area

Branson Homeowners Association Our Newest Venture

We now have the heading and we will add a picture. The next picture we will add will be a picture of golf course.

Click on the Insert Picture **icon**

Navigate to the PowerPoint 2007 folder (or the files that you downloaded) and select the picture "Ireland_Golf" **and then click the** Insert **button**

The picture of the golf course in Ireland will be inserted into the slide, as shown in Figure 8-13.

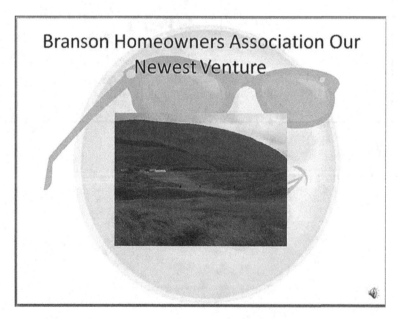

Figure 8-13

As long as the picture is selected, the sizing handles are visible. There is also another handle that is visible. This handle is at the very top and is called the Free Rotate Handle. When you click on this handle and move the mouse to the right or left, the picture will rotate.

Click on the Free Rotate Handle **and rotate the picture slightly to the left as shown in Figure 8-14**

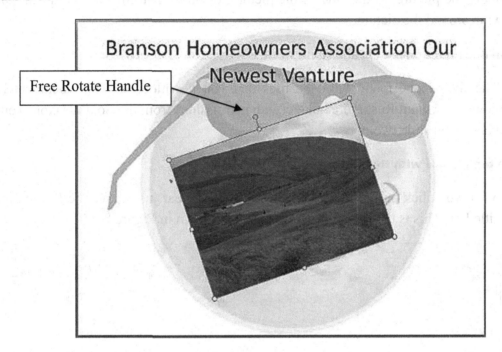

Figure 8-14

Pictures can be rotated to achieve the effect you need for your application.

Note: This photo was used from the PdPhoto royalty free public domain stock photos and can be viewed at their website: http://www.pdphoto.org

Save your work when you are finished

Lesson 8 – 5   Using the Picture Tools

When working with pictures in your presentation you need to decide how you also want the text displayed in relation to your pictures. You can have the text appear under the picture, to the side of the picture, or both. You may also want the text to wrap around the picture.

Open the document called Mount Rushmore

This file can be found with the files that you downloaded. This document has a picture of Mount Rushmore, taken (with permission) from the South Dakota Tourism website, on slide two.

Click on the picture with the mouse

When you click on the picture, a new tab will appear just above the Ribbon on the Title Bar. This is the Picture Tools Tab and is shown in Figure 8-15.

Figure 8-15

This is one of the tabs that are displayed only when needed. It will only be available if the focus is on the picture. The picture has the focus when you click on the picture. Now we will look at the new groups that are on the Format Tab of the Ribbon.

Click on the Format Tab

The first group on the left is the Adjust Group and is shown in Figure 8-16.

Figure 8-16

As you might imagine these are the things that you can adjust on the picture. The brightness will adjust the brightness of the picture. You didn't have to be a rocket scientist to figure that one out. If you click on the Brightness button a list of the preset brightness's will drop down. The live preview will also be available. This means that you can move your mouse down the list and see how each setting will change the picture.

Click on the Brightness **button and then move the mouse slowly down the list to see the effects**

Repeat this for the Contrast **button**

The contrast will adjust the contrast of the picture.

Repeat this for the Recolor **button**

The Recolor will as the name implies change the color of the picture.

Adding pictures can increase the size of the file (presentation) on your computer. Although this is a great asset for your slides, you should realize that the more pictures that are in the presentation the longer it will take to load the presentation. With today's computers this is not a big deal (usually). You can save room on your hard drive and reduce the time it takes to download the pictures by compressing them. This may reduce the resolution slightly and may or may not be visible on the monitor or printer, but it can affect the quality of the picture. This is what the compress Pictures button will do if you click it.

The Change Picture command will let you change this picture for a different one. Clicking this command will bring the Insert Picture dialog box to the screen so you can find a new picture.

The Reset Picture command will remove any formatting that you have made to the picture and restore all of the default settings and formatting.

The next group is the Picture Styles Group and is shown in Figure 8-17.

Figure 8-17

We discussed themes and styles earlier and this group is about the same. It allows you to change the overall style and look of the picture. This group also supports the live preview feature.

Slowly move the mouse pointer over each style and watch the picture on the slide

If you click on the middle down arrow you can try the rest of the available styles. You can also save the formatting in this picture as a new style for you to use later. If you need a recap of themes and styles, you may want to review Chapter six again.

The Picture Shape command will let you use any of the existing Microsoft shapes and totally change the way the picture looks.

Click on the Picture Shape **command and then select the heart which is the basic shapes next to the happy face**

I am sure that you will find a shape that fits into your line of thought for the presentation. The easiest way to get back to the way the picture looked before you started playing is to use the undo button. If you have tried several different shapes you can click on the square in the basic shapes section.

The Picture Border command will let you change the color of the border as well as the style of it. You could use a solid line or dashes or stars or whatever makes it look good.

The Picture Effects command will allow you to change the picture to show different effects. The easiest way to understand this is to try it and see what happens.

Click on the Picture Effects **button**

As you move your mouse down the list of effects a new menu will slide out to the side with more effects under each section. These also have the live preview feature and you can move the mouse over the different icons and watch the results.

Move the mouse over each effect in all of the groups under Picture Effects

The next group is the Arrange Group and is shown in Figure 8-18.

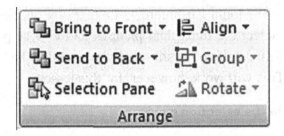

Figure 8-18

We discussed the Bring to front and Send to back in Chapter seven. On slide three we also have some text along with the picture.

Click on the picture in slide three

Alternate between the Bring to front **and** Send to back **commands**

The Selection Pane may be one of the commands that you will not often use. This pane, shown in Figure 8-19, lists all of the objects on the slide and their locations.

Figure 8-19

With the Selection Pane you can show or hide all of the objects on the slide. This includes the picture and the text. You can also change the order they appear on the slide.

175

Click on slide four

Slide four has five pictures of Mount Rushmore and a text box on the slide. We can use the Selection Pane to select the individual pictures and choose perhaps to hide some of them. I know that this slide looks silly and you wouldn't have five pictures all the same on one slide. This will work, however, for this lesson.

Click on each picture mentioned in the selection pane

Click on the name that is in the pane and not on the picture. You can select any picture that you like and if you click on the small icon on the right side of the picture it will toggle the Hide/Show feature for the picture.

If the pictures were different, you might decide that having them in a different order might make your slide even more interesting. If you select a picture and then click on the re-order arrows at the bottom of the pane, the position of the picture will change.

Close the Selection Pane by either clicking on the close button (the small box in the upper right corner with an X in it) or by clicking the Selection Pane command a second time

The Align command will align selected objects in the slide. This can be one slide or several slides.

Select all five pictures on the slide

This can be done by clicking on one picture and holding the CTRL button down as you click on each of the remaining pictures. When you are finished all of the pictures should have the sizing handles and the free rotate handles showing. Figure 8-20 shows the screen.

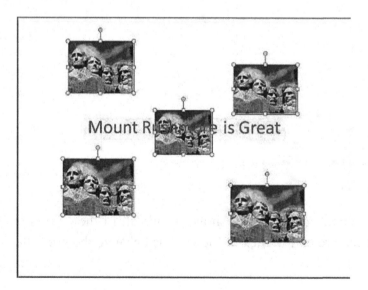

Figure 8-20

Click on the Align command

A drop down menu will slide down and you can choose how you want to align the pictures. This type of align is different from the paragraph align command. If you have only one object selected the align buttons work like you might expect, but with more than one object selected your alignment choices will align the pictures with each other. This sounds strange, so let's give it a try and then you will understand.

Click the Align Left choice

This will cause the pictures to line up on the left side of the picture that is furthest on the left.

Click the Undo button to bring the pictures back to where they were

Click the Center Align button

The center of the pictures will line up with the picture in the center.

Click the Undo button again and then choose the Right Align command

The right side of the pictures all line up with the picture that is furthest on the right.

The top, bottom, and middle align commands all work like the ones you just tried. You might want to click the Show Gridlines choice to make viewing these movements easier.

The Rotate command will do just as it implies. It will rotate the selected object(s) to the left and right. You can also flip the object either vertically or horizontally.

Select any one of the pictures and then move the mouse over each choice in the "Rotate" command

I think my favorite is the Flip Horizontal effect, but that is just me.

The last group on this tab is the Size Group and is shown in Figure 8-21.

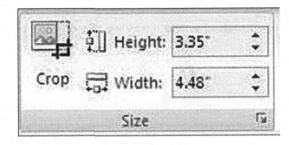

Figure 8-21

The Crop command will allow you to adjust the size of the picture by cutting some of it out. You can choose to make the picture larger, but then you will end up with a section that has nothing in it. We will demonstrate this to you now.

Click on the picture on the top left side and then click on the Crop **command**

You may need to click on the Picture Tools Tab to get the format tab back. The selected picture should now look like Figure 8-22.

Figure 8-22

Move your mouse pointer to the center of the right side of the picture

The mouse pointer will turn into a small T that looks like it is lying on its side. When it changes into this you can click and hold the left mouse button and drag the mouse to make the picture smaller or larger.

Note: When you crop a picture it does not make the entire picture smaller, it cuts part of the picture away. It does not resize the picture which will keep the entire picture intact and make the whole picture smaller. Likewise, if you try to make the picture larger you will only add a blank space on the side or the top.

In contrast, the Height and Width adjustments will increase or decrease the entire picture to make it larger or smaller.

Select one of the pictures and then using the up and down arrows (to the right of the crop command) adjust the height and width of the picture

Close the presentation when you are finished without saving the changes

Lesson 8 – 6 Working with Charts

Another way to make your presentation look professional is to add charts to your slides. In this lesson we will add a chart to our Branson Homeowners Association presentation.

Open the Homeowners Association **presentation**

Press CTRL **and** End **to get to the last slide**

Add a new slide to the presentation

Add the Title

Branson Homeowners Association Projected Profits

Charts are a little more difficult to work with than anything we have done yet. We not only have PowerPoint that we are working with, but we will also be working with Microsoft Excel, a spreadsheet application. When we insert a chart into our slide, we have to establish each value for the columns and rows (the X and the Y axis) or if we use a pie chart we have to establish the value for each section of the pie. In our project we are going to use a simple bar chart to show the projected profit for four years. Are you ready? Here we go!

Click on the Insert Charts **command**

Guess what? You have another choice to make. What type of chart do we want in our presentation? There are about 60-65 different types of charts available for you to choose from. Figure 8-23 show the screen of available charts.

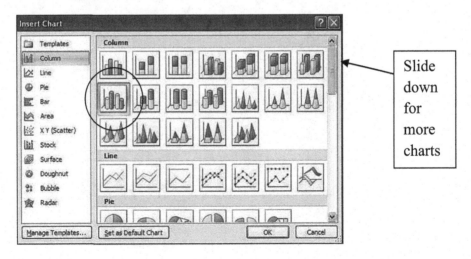

Figure 8-23

For our chart let's choose the Clustered Cylinders (the one outlined in gold in the figure). **Get ready the fun is about to start.**

Click on the Clustered Cylinders **chart and then click** OK

Your screen just went crazy and you are probably wondering why you ever wanted to use PowerPoint.

A chart needs information and you provide that information with an Excel spreadsheet. **Don't panic we are going to get through this.** Your screen should have changed to look like Figure 8-24. You should be able to see both the PowerPoint presentation and the Excel spreadsheet on the screen. You may have to adjust the size of each program to see them both on the screen at one time.

On the top right side of each program there are three buttons. The one with just a line at the bottom is the Minimize button and will cause the screen to shrink down until it is just a small box on the task bar at the bottom of the computer screen. The center button is the Maximize button and will have the program fill the entire screen. If the program does fill the entire screen, clicking this button will make it smaller. If the program does not fill the entire screen the bottom right corner will have several dots in it. Clicking on these dots and dragging the mouse to the left or the right will let you resize the program until it only fills about one-half of the computer monitor. You may have to do this for both PowerPoint and Excel to be able to have both programs on the screen at one time.

If you think about it for a minute, this is pretty cool. These two programs are going to be tied together for our presentation. Both are going to be active parts of our presentation. Just think about it, if we change the values in Excel our chart can also change. This is really cool stuff we are doing.

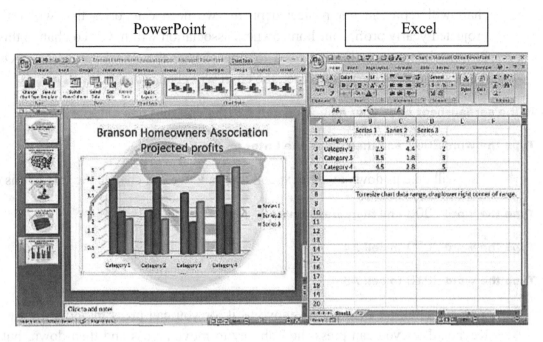

Figure 8-24

Look at the top of the spreadsheet, in the Title bar. It says this is a chart in Microsoft Office PowerPoint and the program is Microsoft Excel. See even Microsoft knows that these two are going to be tied together.

Before we can make any changes we must know which program is the active program. You can switch back and forth between the programs and make any changes needed on either side. However you must make the program active before you can do anything to it. You can tell which program is the active program by looking at the title bar. The title bar will be faded or washed out in the program that is **not** active. The title bar will be darker in the program that **is** active.

If the spreadsheet program is not active, click on the Title bar of the Excel spreadsheet program

On the PowerPoint side of the screen you will notice that there are "categories" below the cylinders which will let us know what the columns represent. The "series" will represent how tall the cylinders will be. Following this same line of thought, you can see where each is represented in the spreadsheet. The categories are on the left and the series are across the top.

181

The categories will need to have names that will make sense to the viewer. Since the chart will represent our projected profits we need categories that will reflect the projected yearly profits our homeowners association will have. To change this in the PowerPoint presentation we need to first change the Excel spreadsheet. Once we make a change in the spreadsheet, the chart will change automatically.

In the spreadsheet click your mouse on cell A2 (the one that has category 1 in it)

Type the word 2007 and then press the Enter on the keyboard

This will take us down to the next cell (A3) and we can change the text in this cell.

Type the word 2008 and then press Enter

Type the word 2009 in cell A4

Type the word 2010 in cell A5

You can click on each cell to move to it, or you can press the down arrow on the Keyboard, or you can press the Tab key to move across and then down, but this is harder than just clicking on the cell with your mouse or using the down arrow key. The change you make will not take effect until you click outside of the cell.

There is something you should have noticed. As you changed the name in the spreadsheet, the category names also changed.

When you have finished this part of the changes, your spreadsheet should look like Figure 8-25.

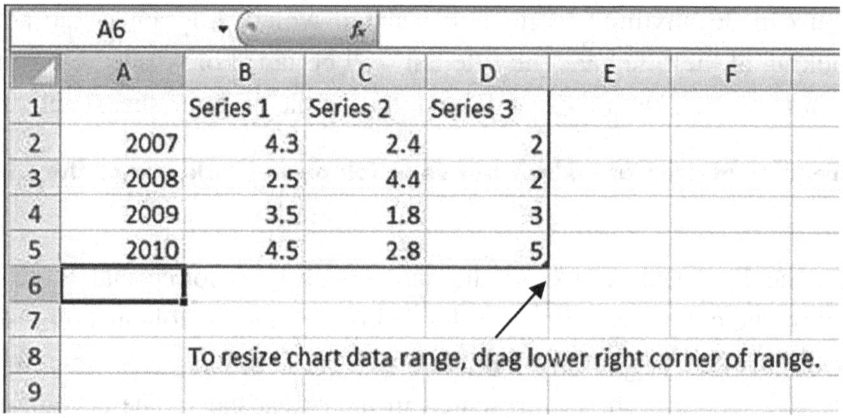

	A	B	C	D	E	F
1		Series 1	Series 2	Series 3		
2	2007	4.3	2.4	2		
3	2008	2.5	4.4	2		
4	2009	3.5	1.8	3		
5	2010	4.5	2.8	5		
6						
7						
8		To resize chart data range, drag lower right corner of range.				
9						

Figure 8-25

The next thing we need to change is the text "Series 1". We will want our chart to show that we are looking at our profits.

Change the text in Cell B1 to be Profit

We will not be needing columns C and D in our chart. If we were looking at more than one thing we would need other references. If we were comparing income and expenses, we would have two columns, one representing each.

Click the lower right corner of the range of cells (where the arrow is pointing in Figure 8-25) and drag the mouse to the left until columns C and D are highlighted in gray, and then release the left mouse button.

The spreadsheet should now look like Figure 8-26.

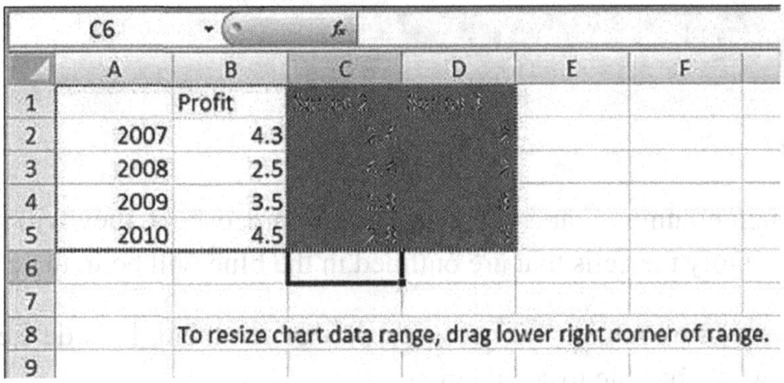

Figure 8-26

You can tell when the mouse is in the correct position because the mouse pointer will turn into a smack black double sided arrow, and it will be at an angle. It will look similar to this: (↖).

When the mouse turns into the double sided arrow, we can click the left mouse button and drag the mouse to the left to move the boundaries of the chart to only include the first column.

Click anywhere outside of the chart area

Columns C and D will no longer be included in the chart and the chart in the PowerPoint slide will only have one set of cylinders in it (See Figure 8-27). Also the blue line will show that columns C and D are not included in the chart area of the spreadsheet.

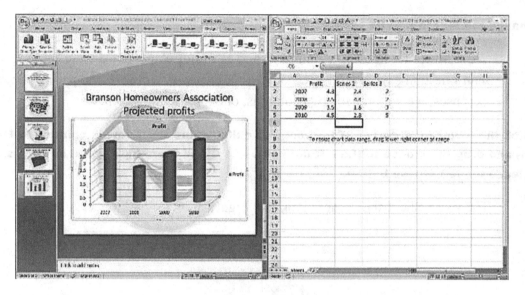

Figure 8-27

Even though columns C and D are still in the spreadsheet, they will not appear in our document. Only the cells that are outlined in the blue will be in the chart.

Just so we don't confuse you because you can see them, let's delete columns C and D so you won't be able to see them.

Click your mouse on the letter C on the Excel spreadsheet and then drag the mouse to the right until column D is also highlighted and then press the delete key on the keyboard

You can tell when the mouse is in the correct place because the mouse pointer will turn into a solid black arrow.

Click the mouse anywhere on the spreadsheet so the columns will no longer be highlighted

That should make the spreadsheet look better, and take a little stress out of your life.

When you are finished your screen should look like Figure 8-28.

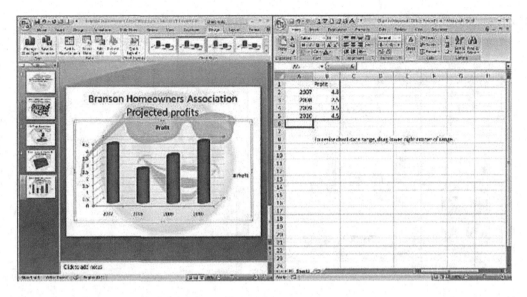

Figure 8-28

We only have a couple of things I want to do to this slide and then we will be finished with it. The next thing I want to do is give the label "Profit" a different name so it will make sense. We want the viewer to know that we are using the U.S. Dollar as our unit of measurement. If we just change the label to say U.S. Dollars it will look like we only had 4.3 dollars as our profit in the first cylinder. Well that won't do. What would look better is if we had the label U.S. Dollars in Millions as the label.

Using your mouse click on cell B1 (the one that has Profit in it) and type

    U.S. Dollars in Millions

Click the mouse in any other cell

Your screen should now look like Figure 8-29.

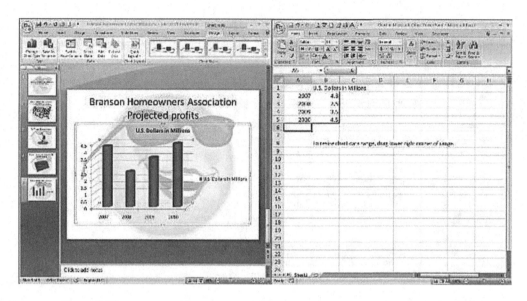

Figure 8-29

The last thing I want to do has nothing to do with how the chart will look or perform. This will just make it a little easier on the eyes when you are looking at the spreadsheet.

Move your mouse to the line that is between the letters B and C on the spreadsheet

When it turns into a small dark plus sign double click the left mouse button

> This will tell Excel to make the column large enough to show all of the words in the column.

> We can now close the Excel spreadsheet program and concentrate on our presentation.

Click the close button on the spreadsheet

> The close button is the small square in the upper right corner of the program. This square has a red X on it.

> Now let's check out our presentation.

> This looks pretty good, but I am not sure that I want to show such a large drop between 2007 and 2008. Let's change the data, so the profits look a little better. Hey this is only make-believe, we can manipulate the data any way we want.

Right click inside the body of the chart

This will bring the shortcut menu for the chart to the screen. The menu is shown in Figure 8-30

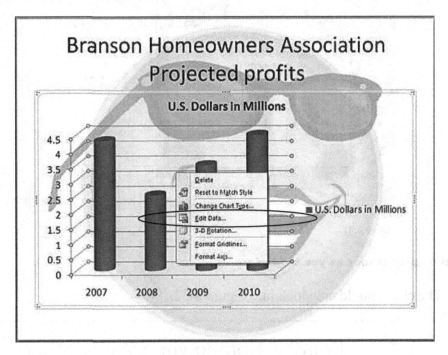

Figure 8-30

Click on Edit Data from the shortcut menu choices

Guess what, the Excel spreadsheet opens back up and we can edit the data in the spreadsheet which will change the data in the chart.

Change cell B2 to: 2.5

Change cell B3 to: 2.7

Change cell B4 to: 3.1

Change cell B5 to: 3.7

If you cannot remember how to change the data inside a cell, go back and check how you changed the headings under the columns.

The spreadsheet and chart should look like Figure 8-31

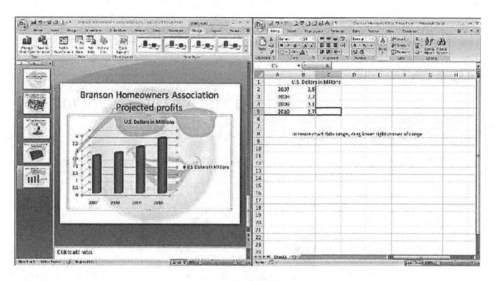

Figure 8-31

Close the spreadsheet as you did before

The chart should now look like Figure 8-32.

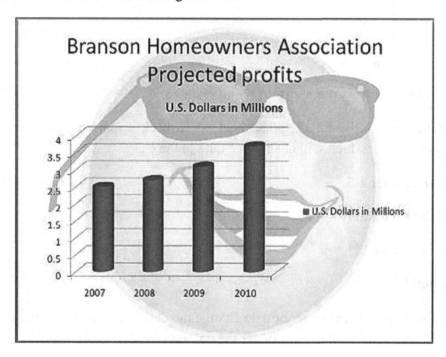

Figure 8-32

On the Design tab you can also change the color of the cylinders, by clicking on one of the chart themes. You can also change the colors by clicking the down arrow next the word "Colors" at the end of the themes.

From the Layout tab, you can insert picture, shapes, and text boxes into the chart area. You can also change how the titles and labels are displayed.

Just for practice, let's insert a shape into out chart.

Click on the Layout tab **and then click on** shapes **in the Illustrations group**

Find the Happy Face **under the Basic Shapes group and click on it**

When you click on the Happy Face, the mouse cursor will change to a plus sign.

Move the plus sign above the legend "Income in Millions" **and click the mouse button**

If the Happy Face is not positioned as shown in Figure 8-33, click on it with the mouse and drag it to the desired location in the chart.

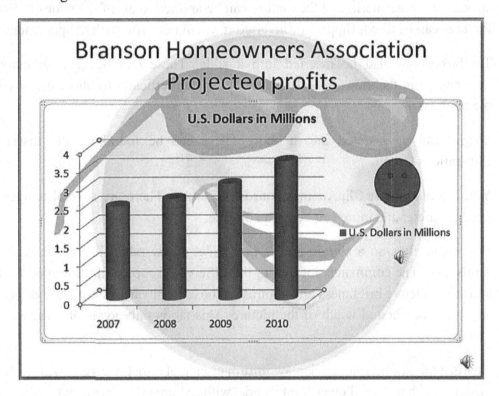

Figure 8-33

Save your work

Chapter Eight   Review

Pictures can be inserted into any slide in your presentation. You can choose from the various choices of slides by clicking the Layout command that is in the Slides Group of the Home Tab. Some choices will allow you to have a caption under the picture and some will allow you to have more than one picture on a slide. Pictures can be inserted into a blank slide by using the Picture command which is in the Illustrations Group of the Insert Tab. Pictures can also be inserted as a background. The transparency can be increased so the picture can appear as a watermark, and the picture can be applied to all of the slides in the presentation. Pictures can be tilted, flipped, or inverted if it will help you with the presentation.

Textboxes can also be inserted into a slide. These can be used to convey additional information to the viewer. Textboxes will expand vertically to allow the words to fit inside the textbox.

Shapes can be inserted into a slide. These can be found on the Insert Tab and the Illustrations Group.

You can also insert ClipArt into your presentation. There are several images that come on your computer and more can be found on-line.

When you click on a picture that is in a slide, a new tab will become visible; the Picture Tools Tab. The commands located in the different groups will allow you to do things like: adjust the picture brightness and contrast, choose from various border styles and shapes, and adjust the height and width of the picture. This tab is only available when you click on the picture.

Charts will allow you to display information that most viewers will understand almost instantly. Charts in PowerPoint work with Microsoft Excel, which is a spreadsheet application. The data is entered into the spreadsheet and Windows will put the information into the PowerPoint Presentation in the form of a chart.

Chapter Eight   Quiz

1) Pictures are the only type of graphics that can be inserted into a slide. **True or False**
2) How do you bring the Format Background Dialog Box to the screen?
3) Textboxes will expand horizontally to allow the typed text to fit into the textbox. **True or False**
4) You can adjust the size of a ClipArt image by click on and dragging it by its _____ _____.
5) To rotate a picture to the left, you would click on the _____ _____ _____ and move the mouse to the left.
6) If you click on a picture that is in a slide, a new tab will appear. What is the name of this tab?
7) The Insert Chart command is found on which tab and group?
8) What program works with PowerPoint to allow you to show a chart in a slide?
9) Once a chart in a spreadsheet, the information displayed cannot be changed. **True or False**
10) If you are working on a chart and the Excel program has been closed, how do you bring the spreadsheet back to the screen?

Chapter Nine   Transitions & Animation

You have probably seen a slide show presentation where the slides fade in and out, or have a way of appearing on the screen in a way that seems so cool that you wish your presentations looked like this. In this chapter we will be changing how your slides appear on the screen. This is called Transitions. We will also be changing how the text and objects appear on the slides. This is called Animation.

If you want your presentations to look like a professional designer created them, you will like this chapter.

Lesson 9 – 1   Slide Transitions

Slide transitions determine how the slide will appear on the screen. PowerPoint 2007 has 56 different effects that you can add using slide transitions. These are totally optional effects and you do not have to include them in the presentation. You could choose to have the slides just appear on the screen. That just wouldn't be any fun at all. Let's add some different effects to our Branson Homeowners presentation.

Open the Homeowners Association presentation

We will start with the first slide and see what we like for this slide. The Animations Tab is where you will find the different transition available.

Click on the Animations Tab

Depending on the available screen width of your monitor, you may see more or less transitions than are shown in Figure 9-1.

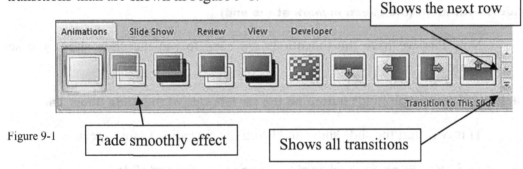

Figure 9-1

One of the great things about selecting an item from this group is the live preview feature that we have talked about in several of the previous lessons. You can let your mouse hover over each transition and see how it will affect your slide.

Move your mouse over to the second transition and watch your slide

The first transition, the one with the gold square around it, is the "no transition". That means that the slide will just appear on the screen when it is time to display it. We will run the entire presentation after we have added a transition to each slide a little later in this lesson.

When you move your mouse to the transition your screen will go black and the slide will fade in. This is a pretty neat effect, so let's add it to the slide.

Click your mouse on the Fade smoothly **transition**

Your screen may flash momentarily and the gold square will move to the second transition to show that it is active.

Click on slide two

Slowly move your mouse over each of the transitions to see what they will be like if you decide to use one of them

After you are finished, click on the Dissolve **transition**

Click on slide three and then click on the first down arrow to see the transitions in the 2nd row

You will have to search for these, but add the following transitions to the slides

Slide 3 – Box Out

Slide 4 – News Flash

Slide 5 – Random **(the question mark at the end)**

Now that we have the transitions we need to run the slide show and see how it works.

Click the "From Beginning" button on the far left side of the Slide Show Tab

This will start the slide show and bring the first slide to the screen.

Press the Spacebar **on the keyboard to move to the second slide**

Repeat this until you have run out of slides and then press the ESC **key on the keyboard**

This is not the only way to advance to the other slides in the presentation. The right arrow will move to the next slide and the left arrow will move to the previous slide. The Page Down button will also move to the next slide and the Page Up button will move to the previous slide. You can also move to the next slide by clicking the left mouse button, if that option is checked on the Animations Tab.

Now that the slide show is finished, let's see what else we can do with the presentation.

Save your work

Lesson 9 – 2   Transitions Sounds

Wouldn't it be great if we could add sounds when we changed to another slide? How about if we could change how fast the slide appeared on the screen? In this lesson we will learn how to do just that.

Open the Homeowners Association presentation if necessary

Click on slide one to make sure it is the active slide

Slide one will be the active slide if you just opened the presentation. If the presentation was still open after the last lesson, slide one might not have been the slide on the screen.

For this lesson we will be using the right side of the Animations Tab. This part of the tab contains the Transitions sound and speed adjustments. It also contains the option to advance to the next slide when the mouse is clicked. The other option under the Advance Slide will be discussed in Chapter 11.

I was thinking that it would be appropriate if there was a round of applause when we started our slide show. Since we cannot count on the people watching the presentation to give us one, let's add that sound to the first slide transition.

Figure 9-2 shows the right side of the Animation Tab.

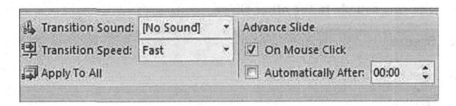

Figure 9-2

Click the down arrow on the Transition sound and choose Applause from the list

A list of the available sounds will drop down and be on the screen and can be seen in Figure 9-3.

Figure 9-3

Now every time this slide comes on the screen we will hear the applause. Again I am making an assumption. You must have a sound card and speakers in your computer to hear the sound effects.

Add the following sounds to the other slides

Slide 2 – Breeze

Slide 3 – Laser

Slide 4 – Voltage

Slide 5 – Cash Register

Now we need to see if this really works.

Run the presentation again and see if the sound effects are there

If you are happy with the choices, we can go on to the next step. Feel free to try the other sound effects and see if you like any of them better than the ones we used. If so, change them.

Click on slide 1 to make it the active slide

What we want now is to change the speed of the transition so that each slide will appear on the screen and allow enough time to pass so that the viewer can see the transition clearer.

On the Animations Tab **click on the down arrow under** Transition speed

A small drop down list will appear with three choices on it. The choices are slow, medium, and fast.

By default the speed is fast. We will change the speed and see if we like the slower speeds better.

Click on the Medium choice

Click on slide two and then change the speed to Slow

Run the presentation and see which speed you like the best

Change any of the speeds on any of the slides to the speed you prefer

The last thing we want to look at is the "Apply to all" button. If you click this button the current slides transitions will be applied to all of the slides in the presentation. Let me say that again. Whatever the transitions of the slide that is showing on the screen will be applied to all of the slides in the presentation if you click the "Apply to all" button. This includes the speed, the sound, and the type of transition. Make sure this is what you want to do before you click this button.

Save your changes

Lesson 9 – 3   Animations

The Animation is a little different than the transition. The transition tells how the entire slide is brought onto the screen while the animation tells how the text and objects are brought onto the slide.

Open the Homeowners Association **presentation if necessary**

Click on slide one to make sure it is the active slide

Slide one will be the active slide if you just opened the presentation. If the presentation was still open after the last lesson, slide one might not have been the slide on the screen.

Click inside the text that is on the slide

This will allow us to add animation to the text. Now we can click on the drop down arrow next to Animate and choose one of the default choices or add a custom animation to the text.

Click on Fade

You will get a preview of the effect when you move the mouse over the effect and also when you click on the effect to apply it to the slide.

Add the Wipe **animation to the text in slide 2**

Add the Fly in **animation to the text in slide 3**

We will use the Custom Animation to add an effect to slides 4 and 5

Click on slide 4 and then click on Custom Animation

The Custom Animation pane will appear on the right side of the screen and we can choose one of the effects for each category listed in the effects drop down menu. We do not have to use all of the available choices but we can if we want. We can choose an effect for the object when it (the text or picture or chart or table) enters the slide, you can add an effect to show emphasis, you can add an effect when the object leaves the slide, or we can add a motion path the object and cause it to move.

This is really confusing, so I think we should apply some effects and then you will understand how they work.

Click on the text in the slide

Now we will be able to add an effect. We could also have clicked on the picture in the slide and added an effect for that also. First we will add an effect to the text and then we will consider the picture.

In the Custom Animation pane click on Add Effect

The first effect that we will add is upon the entrance of the text.

Move the mouse to the Entrance **choice and then click on the** Checkerboard

When you move your mouse to the Entrance choice another menu will slide out to the side and then you can click on the Checkerboard choice. Figure 9-4 shows the menu.

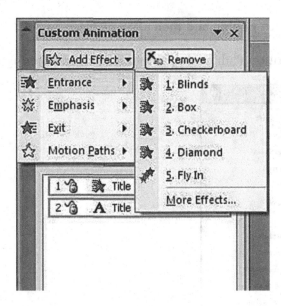

Figure 9-4

We will leave the insertion point (the flashing vertical line) inside the textbox and add another effect to the text. This effect will show emphasis on the text.

In the Custom Animation pane click on Add Effect

Move the mouse to the Emphasis **choice and then click on the** Change Font **choice**

We now have two effects added to the text. You will get to see how these work in a few minutes. Before we start trying the effects out let's add another effect.

199

In the Custom Animation pane click on Add Effect

Move the mouse to the Exit **choice and then click on "More effects" choice**

> This will bring a dialog box to the screen and we can choose the effect we want from there.

Move the scroll bar down until the Exciting **group becomes visible and click on** Curve down **from the list of choices**

> This will cause the text to curve out of the slide and exit. The last thing we want to do to this slide is have the picture move to a new location in the slide. We do this by clicking on the picture and then adding the motion path effect to the picture.

Click on the picture and then in the Custom Animation pane click on Add Effect

Move the mouse to the Motion path **choice and then click on "More effects" choice**

> This will bring the Add Motion Path Dialog box to the screen (See Figure 9-5).

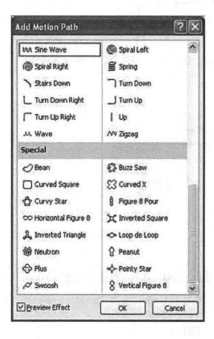

Figure 9-5

Scroll down to the Special **section and then click on** Buzz Saw

> This will give us enough to work with. Let's run the presentation and then you will probably understand what we have done and how each animation affects our presentation.

On the Slide Show Tab and in the Start Slide Show group, click on "From Beginning"

There will be several things that you will notice. The first thing you will notice is that the sound starts before the slide is displayed on the screen. The second thing you will notice is that the text never appeared on the screen.

Click the mouse or press the Spacebar **to advance the presentation**

Clicking the mouse did not bring the second slide to the screen; it brought the next effect to the screen. This is the process that will continue throughout the slide show. Each mouse click will move one step, not necessarily one slide like it did before we added the effects.

Keep going until you have seen slide 2 and 3 but do not go on to slide 4 yet

Slide 4 is the slide that has several effects in it. We want to go through this slide a little slower to make sure you see all of the effects and understand them.

First the picture will spin onto the slide. Second the text will enter. Third the font will change. Forth the text will exit the slide. Fifth the picture will go into motion. Each of these effects will happen as you click the mouse or however you decide to progress to the next step.

Click the mouse and move on through the effects and the last slide

Slide 5 did not have any effects attached to it. We need to fix that.

Click on slide 5 and add a motion path to the happy face, have it bounce off the screen

I am hoping that you will figure this out all by yourself and not have to read the next paragraph that tells you how to get this added to the slide.

Click on the happy face

Click on Add Effect **on the Custom Animation pane**

Move the mouse over to Exit

Click More effects

Choose Bounce **from the** Exciting **group**

Click OK

Now at the very end of the presentation the happy face should bounce off of the slide.

Run the presentation and see if this is what happens

I hope you noticed something. Our presentation doesn't flow quite right. It seems that it would be better if the words appeared first and not the pictures. The pictures should appear after the words are already on the screen. Let's see if we can fix this.

Click on slide 2 (make sure the Custom Animation pane is on the right side of the screen)

This is where we will make changes to the animation part of our slide. First we will remove the animation from the text so that it will be on the screen when the slide appears. Next we will add the same animation that was on the text to the map and the textbox and the arrow pointing to where Branson is located. I know it sound like a lot but it really isn't that bad.

Click inside the text at the top of the slide

The Custom Animation pane will show the effects that are tied to this object (the text). Figure 9-6 shows the Custom Animation pane.

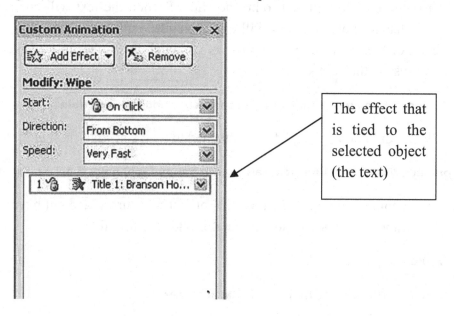

Figure 9-6

You can see that the text has one effect tied to it and this is where we will remove the effect.

Click the down arrow and then click "Remove" at the bottom of the list

The effect is now gone and is removed from the pane.

Now we have to add the effect to the pictures on the slide. This will involve doing something that we have not done before. However, I have faith in you and am pretty sure that you can do this.

Click the mouse on the map

Hold the CTRL **key down and click on the words** Branson Missouri

Continue to hold the CTRL **key down and click on the** arrow showing the location

Release the CTRL key

Holding the CTRL key down will allow you to select more than one object. Every object that you click on while holding the CTRL key down will also be selected and you can treat them as a group and add the effect to all three objects. This will allow the effect to happen to all three objects at the same time. This will be better than having to click the mouse three times, one for the map, one for the name, and one for the arrow. One mouse click will perform the effect for all three objects at once.

Your slide should look like Figure 9-7.

Figure 9-7

For slide two we used the Wipe animation. That is the one we will use for the three objects that we have selected.

Click on Add Effect

Move the mouse down to Entrance **and then click on "**More effects**"**

Find Wipe **in the** Basic **section and click on it**

Now that you have the animation in place, let's see what it looks like.

Run the presentation

Change the animation on slide 3 to have the text on the screen and the effect on the picture of the golfer

Just to save you some time I will remind you of the animation on the slide (Slide three had the Fly in animation).

On slide 4 remove the Buzz Saw effect from the picture and add an Entrance effect to the picture. Use the Grow and Turn **entrance**

This can be found in the "More effects" section.

Run your presentation and see how it looks

Before we leave this lesson there is one other thing that you might enjoy. Let's add something to the last slide.

Click on slide 5 and the click on the happy face

Click the down arrow where it says "Smiley face 5" **and then click on** Timing

A new dialog box will come onto the screen and is shown in Figure 9-8.

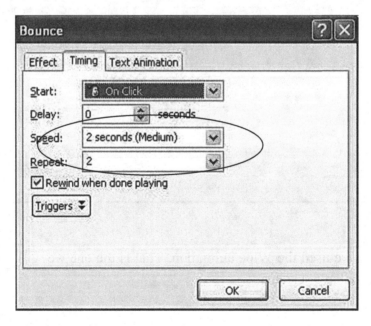

Figure 9-8

Make the 2 changes that are circled in Figure 9-8

Click the OK **button**

Now run the slide show again

The happy face on the last slide should repeat the effect two times and then rewind. This "rewind" will cause the happy face to come back onto the screen.

Adding transitions and animation to your show will make your presentations a lot more fun for you and the viewer. Hopefully this will also make your presentations stand out over your competition's presentations.

Save your changes

Chapter Nine   Review

Transitions and Animation will give your presentations that "little extra" that will let them stand out from everyone else's presentation.

Transitions affect how your slides appear on the screen, while Animation affects how the text and objects appear on the slides.

Both Animation and Transitions are found on the Animations Tab of the Ribbon.

You can also add sounds as part of the transitions, and change the speed of the transition.

Animation will allow you to change how the text and objects appear on or exit the slides.

Chapter Nine   Quiz

1) Fill in the blank. _____ affect how the slide appears on the screen, while _____ affects how the text and objects appear on the slide.
2) Transitions use the live preview feature. **True or False**
3) After you add transitions, the presentation will run without the user pressing any keys or clicking the mouse. **True or False**
4) In what group and tab on the Ribbon will you find the command to add the applause sound to a slide?
5) Before you can add animation to a slide, you must first add a transition to a slide. **True or False**
6) If you add an animation to an object when it enters a slide, you cannot add an animation to it when it exits a slide. **True or False**

Chapter Ten   Multimedia

There may be times when having a little something extra in your presentation may make the difference between a pay raise and staying where you are in the company. Of course, you might just think this is really cool and you should learn it. Having sound effects as the slides were transitioned was great, but what about having an audio clip inside your presentation. That would be super cool.

This chapter will explain how you can add audio and video clips into your presentation.

Lesson 10 – 1   Adding Audio Clips

There are several web sites that will let you download audio clips at no charge. One of these is www.wavplanet.com. There are audio files beyond your imagination at this site and if they don't have what you are looking for, there are links to other sites. Some of these other sites charge for the clips and some do not.

Open the Homeowners Association **presentation**

Click on the last slide in the presentation

We will add a sound clip to this slide and be able to play it with the click of the mouse. On the Insert Tab is the Media Clips Group (See Figure 10-1).

Figure 10-1

Click the down arrow under Sound **and then click "Sound from file"**

The Insert Sound Dialog box will pop onto the screen. This is where you tell PowerPoint where the audio file is located. **Note:** Clicking the top half of the Sound button would also have brought the dialog box to the screen.

Navigate to the files that you downloaded or the PowerPoint 2007 folder if you copied the files to this folder

Select the Jerrersons.mp3 **file and then click the** OK **button**

I bet you guessed what was going to happen. Another box came to the screen. This time PowerPoint wants to know if the sound should start automatically when the slide is brought to the screen or if it is to be played when you click the mouse on the icon. We only want the sound to play when we click the icon so let's choose the when clicked option.

Click the "When Clicked" button

Figure 10-2 shows the new dialog / information box.

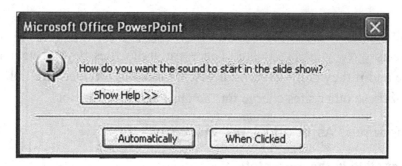

Figure 10-2

The little speaker will be right in the center of the chart. Well that's not good. I suppose we will have to move it.

Move your mouse over to the center of the speaker and when it turns into a plus sign click and hold the left mouse button down and drag it over to somewhere below the U.S. Dollars in millions that is below the happy face then release the mouse button

Now when we run the presentation and before we end the slide show we should be able to click on the speaker and play the theme song from the Jefferson's TV show.

Note: If you do not have a sound card and speakers in your computer you will not hear the music.

Run the presentation and after the happy face comes back onto the screen click the speaker

Save your changes

Lesson 10 – 2   Adding Video Clips

If you thought that the last lesson was pretty cool, wait until you see this one. We can also add a video clip to our slide show. We will use one of the free video clips for our slide show. This file is included in the downloaded files.

Open the Homeowners Association **presentation**

Click on the last slide in the presentation

Let's add a new slide just for the video clip.

Add a new slide after slide 5 and make it a blank slide

We will add a video clip to this slide and be able to play it with the click of the mouse. On the Insert Tab is the Media Clips Group (See Figure 10-3).

Figure 10-3

Click the down arrow under Movie **and then click** "Movie from file"

The Insert movie Dialog box will pop onto the screen. This is where you tell PowerPoint where the video file is located. **Note:** Clicking the top half of the Movie button would also have brought the dialog box to the screen.

Navigate to the files that you downloaded

Select the Boat.mpeg **file and then click the** OK **button**

I bet you guessed what was going to happen. Another box came to the screen. This time PowerPoint wants to know if the movie should start automatically when the slide is brought to the screen or if it is to be played when you click the mouse on the icon. We only want the video to play when we click the icon so let's choose the when clicked option.

Click the "When Clicked" button

Using the sizing handles, increase the size of the video screen until it covers about 2/3 of the slide

Save the changes and then run the presentation

Chapter Ten   Review

You can insert audio and video clips into your presentations. These commands are found in the Media group that is on the Insert Tab.

Make sure you have permission to use any clips that you download (they belong to someone).

You can choose to have the clips run automatically or when you click on the icon with the mouse.

Chapter Ten   Quiz

1) Anything found on the internet is free and can be used by anyone who desires to use it. **True or False**
2) What will happen if you click on the top half of the sound command?
3) You can choose to have audio clips play automatically. **True or False**
4) Video clips must start with a mouse click. **True or False**

Chapter Eleven   Presenting the Presentation

Now that we have our presentation finished, you need to think about how you are going to present your presentation to the viewer. This could be on a computer or on a projector that shows the slide show on a large screen. This could also be a self-running presentation and you are not even there when it runs. Perhaps it would be better to print out your presentation and hand out a paper copy to everyone.

These are the types of things that you need to take into consideration when you think about showing your presentation. We will discuss these kinds of things in this chapter as well as a few other things you might come in handy as you make your presentation.

Lesson 11 – 1   Delivering on a Computer

We are not going to go over all of the things that we have previously mentioned about how to deliver your presentation on a computer. You have run your presentations several times already in this book. Instead we will cover a few things that you might not know yet.

First I want to make sure you know the different ways to move between slides in the presentation. Table 11-1 shows some of the keystrokes that will help you in the presentation.

To Do This	**Press This**
Advance presentation to the next slide or advance to the next effect	Pressing any of these keys will advance: Enter, Spacebar, the right arrow, Page-Down, or click the left mouse button
Go backwards to the previous slide (or previous effect)	Pressing any of these keys will go backwards: The Up arrow, the Left arrow, or Page-Up
Move to specific slide	Enter the slide number and then press Enter
Go back and forth between the presentation and a black screen	Press the B key
Go back and forth between the presentation and a white screen	Press the W key
Show or Hide the mouse pointer	Press the letter A or the equals key
Change the mouse pointer to a Pen	Press the CTRL key and the P key together then release both
Change the mouse pointer back to an arrow	Press the CTRL key and the AP key together then release both or press the ESC key

Erase all of the pen strokes you added to the slide during the presentation	Press the E key (or choose Discard from the dialog box when you exit)
Stop or end the slide show	Press the ESC key

Table 11-1

Some of the things mentioned in the table you have not seen yet. Let's take a closer look at them.

Open the Homeowners Association **presentation**

Move through the presentation until all of slide 2 is on the screen (including the effects)

As you are moving up through the presentation, getting to slide 2, try using several of the methods mentioned in the table.

Move backwards to slide1

Again try using different methods to get to slide 1.

Advance to the end of slide 4 (when the picture is on the screen)

Move back to slide 2 by pressing the number 2 on the keyboard and then press the Enter **key**

That should have moved you to the beginning of slide 2. Advance one more step to show the map on the screen.

Change the mouse pointer to a pen

This is done by pressing and holding the CTRL key and then pressing the P key and releasing both keys.

Draw a circle at the point of the arrow showing where Branson is located (See Figure 11-1)

Figure 11-1

Erase your drawing

If you need to pause the slide show and you don't want the slide to remain on the screen you can have the screen go black or white depending on which key you press.

Go to a black screen by pressing the B key and then bring the slide back to the screen by pressing the B key a second time

Repeat this only this time use the W key to go to a white screen

This might be useful if you had to pause to answer questions and you would prefer not to have the slide visible during this time.

Hiding the mouse pointer is not a big deal unless it interferes with the presentation. This is probably not something that you will use very often, but it is there if you need it.

Lesson 11 – 2   Rehearsing Timings

This doesn't seem like a lesson that you would be interested in, but what if you were giving your presentation in front of a large group and you weren't in a position to click the mouse when you needed to change slides. We could have PowerPoint change the slides for us automatically at preset intervals.

There are a few things you should remember when presenting your slide show. They are: If the slide is on the screen to long the viewer will lose interest in it. On the other hand if the slide is not on the screen long enough, the viewer will not have time to read everything that is on the slide.

Open the Homeowners Association **presentation**

In this lesson, and the next, we will be using the Set Up Group on the Slide Show Tab of the Ribbon. This group is shown in Figure 11-2.

Figure 11-2

We are going to be looking at the Rehearse Timing command in this lesson. This command will start the slide show and you can practice saying everything that you need to say and when you change the slide a timer will keep tract of the time that the slide was on the screen. If the "Use rehearsed timings" option is checked, PowerPoint will use this recorded time to know when to change the slides.

First we have to decide exactly what we are going to say at every slide. This is not a bad as it sounds, since you have to know what you are going to say anyway. I can't imagine that you would go into a presentation and not know what you were planning on saying. Either way we have to rehearse what we will be saying. Normally, you would run the presentation several times just to make sure you covered everything you should say. In this situation I have provided you with a script you can practice with.

Start slide show

Thank you for all of the applause

Thank you very much

(Click the Mouse)

I want to welcome all of you to the annual meeting of the Branson Homeowners Association Goes Across America.

(Click the mouse)

As you know our home office is located in Branson, Missouri. For the new members and visitors (Click the Mouse)

Branson Missouri is located in the Southwest part of Missouri almost on the Arkansas border.

(Click the mouse)

(Click the mouse)

One of the may fun things you can do at our property, besides the swimming and enjoying all of the music shows, is Golf. All of our resorts have a private on-site golf course for your enjoyment.

(Click the mouse)

(Click the mouse)

Today we want to talk about our newest venture

(Click the mouse)

(Click the mouse)

(Click the mouse)

This is the site for our newest venture. There are acres and acres of rolling hills which will be one of the premier golf resorts in the world.

You might ask how we can provide yet another golf resort to our members. Here is the answer.

(Click the mouse)

Our profits have steadily increased over the past 4 years, and they are projected to increase over the next four years. Now we are very HAPPY (Click the mouse) campers

(Click on music) (Do a little dance and get a laugh) We have worked very hard and now we have Moved on up. (Let the music finish)

(Click the mouse)

This has nothing to do with our meeting tonight I just thought that you would enjoy it.

(Click on the Video slide)

When it is finished, click the mouse and the presentation is over.

This is the script that you can practice with for the timers.

Before we click the mouse on the Rehearse Timings command get your script ready. The timer will start when you click the command.

Click on the Rehearse Timings command

Recite the script as you think that a person in front of hundreds of people would say it

Follow the directions and click the mouse when you need to bring the different effects and slides to the screen

When the last slide exits the screen the timer will stop and you will get a new dialog box asking if you want to keep these timers and use them in the presentation.

Choose yes that you want to use these timers

This process can be repeated as many times as necessary until you are happy with the results.

When you agree to use the recorded settings a new screen will jump onto the monitor. This screen will show each slide and the time that it will be on the screen before the next slide replaces it. Your screen will not look exactly like the one in Figure 11-3, but this will give you an idea of what this screen will look like.

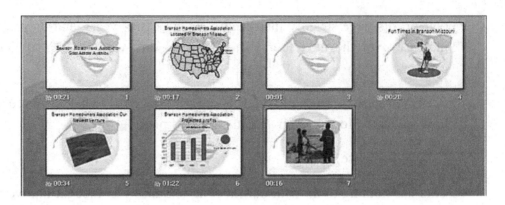

Figure 11-3

If you are wondering what this screen is and where it came from, click on the View Tab and you will notice that the Slide Sorter View is being displayed. To go back to the normal view all we have to do is click on Normal View.

Click on Normal View **on the View Tab**

Run the slide show and talk through the presentation and watch the slides advance automatically. Isn't this great?

Play with this and try changing the settings as you work through the slide presentation. When you are finished:

Save the changes

Lesson 11 – 3   Self Playing Presentation

In the last lesson we had a lot of fun dancing and preparing for the slide show to be presented to the membership of our fictitious organization. Well I had fun doing all of that stuff anyway. In this lesson we are going to pretend that the presentation is going to be running at a kiosk and we are not even going to be there. This will be a totally automated presentation.

The first thing we need to consider is whether we want sound to be a part of the presentation. We might not need sound with every presentation, but it would make sense for ours to have sound. Besides if it didn't need sound we wouldn't have much of a lesson. To have sound you must have a microphone attached to your computer, or have a built-in microphone. Having a built-in microphone is very popular now and you may have one, if not you will need to plug one into the small jack that has Microphone on it.

Click the Record Narration **button**

The Record Narration Dialog box will come to the screen as shown in Figure 11-4.

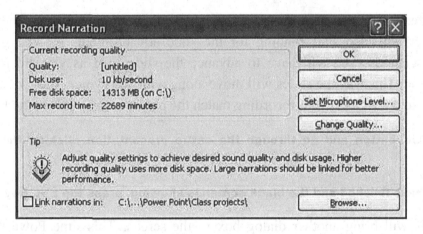

Figure 11-4

The next thing we will want to do is set the Microphone level. This will allow us to see if there is a microphone connected to our computer and if it is working correctly.

Click the Set Microphone Level **button**

This will bring yet another dialog box to the screen. This is shown in Figure 11-5.

Figure 11-5

I took this screen capture when I was speaking into the microphone so you could see what it will look like when you talk into the microphone. If you do not have the green bar when you speak, you may not the microphone plugged into the correct "little hole" on the sound card.

If it appears that the microphone is working and the green bar shows when you speak, click the OK **button**

Note: When you click the next OK button, PowerPoint will start the recorder. The timer will also start running for the slide advancement. As you go through the presentation, you will have to advance the slides just as you did in the previous lesson. This way the slides will move along with your voice. Can you imagine if you had to try to make your recording match the previous timers from the last lesson?

Click the OK **button and go through the entire presentation clicking the mouse when necessary**

When you have finished and the black screen is showing, press the ESC **key**

This will bring another dialog box to the screen. This time PowerPoint wants to know if it should save the new timings for the presentation. If you click the "Don't save" option, the original timings from the last lesson will still be there and your voice will probably not match the slide changes.

Click on the Save **choice**

Run the presentation and see how it looks

As you can see, there are a few little things that were not quite right. The sound effects on the slide transitions can get in the way of your voice and things seem to be off just a little. If you were really making this a self running presentation, you would more than likely not use the sound effects. Also you will notice that the video clip did not start by itself. We clicked on it, but it didn't start by itself. That click is not part of the presentation and is not recorded as a click on the video. If we were going to use the video we would set is to start when the slide is loaded and not with a mouse click.

Remove the sound effects from the slide transitions

To do this, click on each slide and then click on the Animation Tab and then click on the down arrow next to Transition Sound and choose No sound from the list.

Change the Video to run as soon as the slide is loaded and not when clicked

To do this you first have to click on the screen that is inside the slide where the video is going to be played. A new Tab on the Ribbon will become visible, remember that we said there were some tabs that would only be visible when they could be used; well this is one of them. The part we are interested in is the Movie Options Group. This group is shown in Figure 11-6.

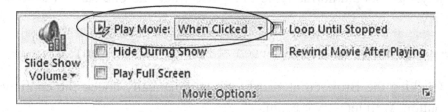

Figure 11-6

Change the Play Movie setting from when clicked to Automatic

The presentation worked just fine other than the things that we just changed, the sound effects and the video clip, but it only ran one time and then quit. At a kiosk we would not be there to start it every couple of minutes. What can we do?

As usual, Microsoft is way ahead of us, and already planned for this. There is a way to have the slide show restart when it finishes.

Click on the Setup Slide Show **command that is located in the Setup group of the Slide Show Tab**

This will bring the Setup Slide Show Dialog box to the screen (See Figure 11-7).

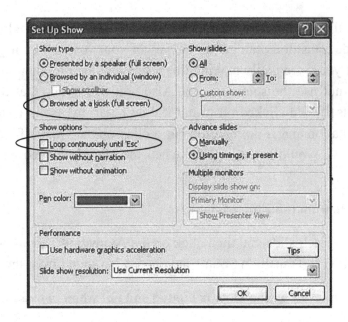

Figure 11-7

There are two things that we need to be concerned with in this dialog box. The first is that the show will be browsed at a kiosk. The second is that we need to continuously loop the slide show until we press the ESC key on the keyboard. The keyboard should normally be in a place where no one can get to it to press the ESC key.

Click the radio button next to "Browsed at a kiosk (full screen)"

Clicking the Browsed at Kiosk will automatically check the continuous loop radio button. Well let's try it again.

We have made several changes to the presentation and we will need to re-record the narration to make it sound correct.

Re-record the narration and then run the slide show

Save your changes when you are finished.

Chapter Eleven Review

When you are presenting your presentation, there are several keys that you can use to move through your presentation. Make sure that you are familiar with Table 11-1.

You can rehearse your presentation and have PowerPoint set the timings to automatically advance the slides for you. These settings can be saved or changed as necessary.

You can set your presentation to be a self-playing presentation by setting it up as a presentation playing at a kiosk. You can record your voice as well as your mouse clicks.

Chapter Eleven   Quiz

1) If you want to pause your presentation and show a black screen, what would you press?
2) What can you press to move backwards one slide?
3) What do you press to change the mouse pointer into a pen?
4) The Rehearse Timings Command is found on which Tab and Group?
5) Which view will show you how long a slide will be displayed on the screen?
6) How can you tell if your microphone's volume is appropriate?
7) If you want your presentation to play by without you being there what must you click in the Setup Slideshow Dialog Box?

Chapter Twelve   Protecting your Presentation

There may be times when you don't want just anyone to view you presentations. With PowerPoint you can password protect presentations. If someone does not know the password, they will not be able to open the presentation. If your job, or pay raise, depends on having the perfect presentation to land a new client this would be a very good thing.

Lesson 12 – 1   Creating a Password to open the Presentation

Open the Homeowners Association presentation if necessary

From this point we can add a password to open the presentation, or we can add the password when we save the presentation. If we choose the second option, we can also add a password to modify the presentation.

First we will see how to add a password to open the presentation.

Click the Office button and choose Prepare from the menu

There are several choices available (see Figure 12-1). We are interested in encrypting our presentation to protect its contents.

At this point, I need to stress the importance of keeping your passwords in a safe place and choosing passwords that not just anyone can figure out.

First, if you forget the password, there is no getting it back. **You will not be able to open your presentation. Put your passwords in a SAFE PLACE!** (And one that you will able to find)

Second, if you don't want anyone to figure it out, you need to use a password that is a mixture of numbers and letters. The password can be up to 255 characters.

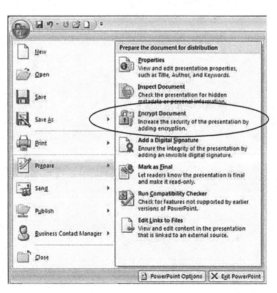

Figure 12-1

Click on the Encrypt Document

When you click on Encrypt Document, another dialog box will appear on the screen. This box is shown in Figure 12-2.

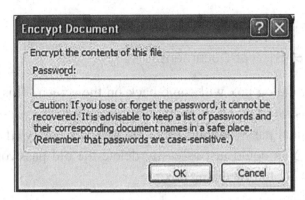

Figure 12-2

What you are going to do is enter a password that will be used to open the presentation. The Presentation will not open without the password you put in the textbox.

In the textbox type 1208 **and then click** OK

A second dialog box that looks almost identical to this one will pop onto the screen. You must re-enter the same password in this box to make sure you put the first one in correctly. All passwords are case sensitive. That means that bill and Bill are completely different. You must enter the second password exactly as you entered the first password.

In the textbox type 1208 **and then click** OK

The dialog boxes will go away and everything is set. The next time you try to open the Presentation you will be prompted to enter the password.

If you try to open the document with the wrong password, you will get an error message stating that you have entered an invalid password.

Save the presentation and then close it

Open the presentation

Instead of the presentation opening, you will get a dialog box asking you to enter the password.

Using the keyboard enter 1208 **and then press the** Enter **key**

The presentation will now open.

Now the only other thing to discuss is how you remove the password if you change your mind and don't want the encryption for the presentation.

Click the Office button and select Prepare

From the slide out menu select Encrypt Document

The encrypt document dialog box will come back on the screen. This time there are four asterisks in the password text box. If you want to remove the password, you will have to delete the original password and leave the text box empty (If you wanted to change the password. You could just as easily delete the old password and enter a new one). See Figure 12-4 for the dialog box.

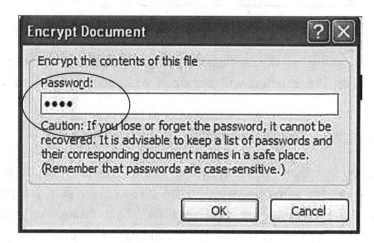

Figure 12-4

Press the Delete **button until the password is gone, and then click** OK

The file is no longer password protected and anyone can open it.

Repeat these steps and put the password 1208 back in

Save this presentation and then close it

Lesson 12 – 2   Creating a Password to Edit a Presentation

Having a password that will keep people from opening a presentation is great, but what if you want to keep someone from editing one of your presentations after it is open? PowerPoint has come up with a way for us to accomplish this also.

Open the Homeowners Association **presentation**

You are prompted for the password. Enter the password (1208) and click OK. The document will open and you can work on it as necessary.

Our needs have changed and we want to protect what is in the presentation from being changed.

From the Office Button click Save As

This will bring the Save As Dialog box to the screen.

Click the Tools **Button on the bottom left of the dialog box**

This is shown in Figure 12-5.

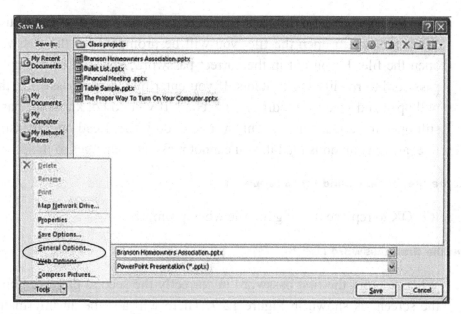

Figure 12-5

Click on General Options

A new dialog box comes to the screen, the General Options dialog box. From here we can set the password for opening the presentation and we can set a password for modifying the Presentation. Figure 12-6 shows the General Options Dialog Box.

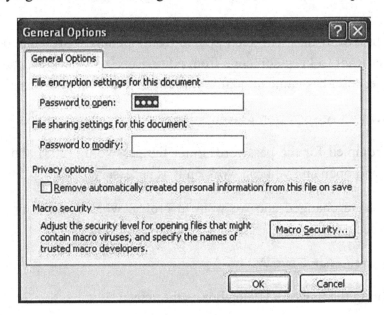

Figure 12-6

Add the password 3423 **to the modify password**

Now we have added a different password to modify the contents of the presentation. When you try to open the file, you will be prompted for the password required to open the file. If you put in the correct password, you will be prompted to enter the password to modify the contents. If you enter the correct password, the presentation will open and you can modify it as needed. If you do not know the password you can still open the document, but only as a read only file. Read only is just what it sounds like, all you can do is read it, you cannot make any changes to it.

Save the file, close it, and try to reopen it

It is OK to replace the original file when prompted.

Put in the first password (1208**)**

When you put the first password in correctly the second password box will come to the screen, as shown in Figure 12-7. There will also be an information box telling you that you can cancel opening the document by pressing the ESC key.

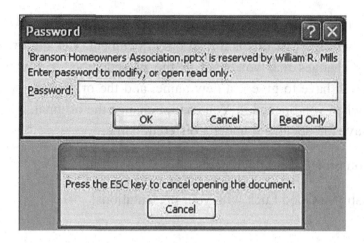

Figure 12-7

You might notice that the file is reserved and who reserved it. If you enter the next password (3423) the document will open. If you do not know the password and you want to see what is in the document, you will have to click the Read Only button.

Click the Read Only **button**

When you click this, the document will open, but no changes can be made to it.

Notice Figure 12-8 and see what a read only document looks like.

Figure 12-8

The document itself looks like any other PowerPoint presentation. At the top, in the Title Bar, you will see that this is a read only document. **Let me explain this**, you can make any changes that you want to the presentation. If you try to save the changes, you will have to give it a new name, and the original presentation will be just as it was when you opened it. You cannot change the original, but you can make changes and save the document under a different name.

That is how you protect your documents.

With this I wish you Good Luck with your presentations!

Chapter Twelve   Review

In this chapter you have learned how to password protect your presentation. If you encrypt your presentation and you forget or lose the password, you cannot open the presentation. There are no reminders or hints to help you.

You can encrypt a document by choosing Prepare and then Encrypt Document after clicking the Office Button, or when you save the presentation using the save as option.

You must enter the password twice when you encrypt the document, just to make sure you entered it correctly.

You can also require a password to edit or modify your presentation. This is done from the Save As Dialog Box, under tools and then general options. If you open the presentation as a Read Only file, you cannot edit the original file. To save any changes made will require you to save the presentation with a new name.

Chapter Twelve   Quiz

1) The only way to have a password to keep someone from opening your presentation is to use the Encrypt command that is under the Prepare option when you click the Office Button. **True or False**
2) When you first encrypt a document, the password must be entered into two different dialog boxes. **True or False**
3) The following two passwords are the same: DAVE and Dave. **True or False**
4) What dialog box is showing on the screen when you set the password to modify a document?